DNA REPLICATION

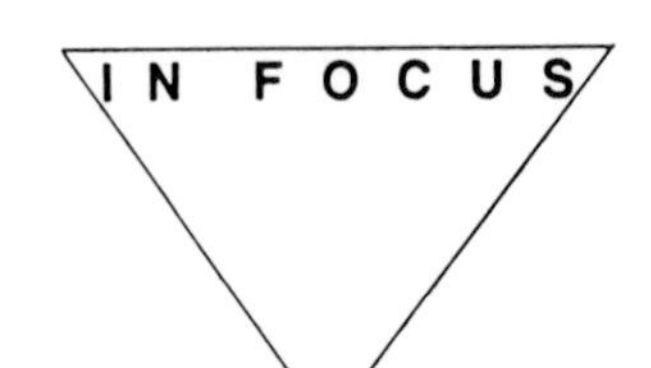

Titles published in the series:

***Antigen-presenting Cells**
DNA Replication
***Complement**
Enzyme Kinetics
Gene Structure and Transcription
Genetic Engineering
***Immune Recognition**
***B Lymphocytes**
***Lymphokines**
Membrane Structure and Function
Molecular Basis of Inherited Disease
Protein Engineering
Regulation of Enzyme Activity

***Published in association with the British Society for Immunology.**

Series editors

David Rickwood

Department of Biology, University of Essex, Wivenhoe Park, Colchester, Essex CO4 3SQ, UK

David Male

Institute of Psychiatry, De Crespigny Park, Denmark Hill, London SE5 8AF, UK

DNA REPLICATION

Roger L.P.Adams

Senior Lecturer, Department of Biochemistry, University of Glasgow,
Glasgow G12 8QQ, UK

Oxford University Press
Walton Street, Oxford OX2 6DP

Oxford is a trade mark of Oxford University Press

Published in the United States
by Oxford University Press, New York

British Library Cataloguing in Publication Data
Adams, Roger L. P.
DNA replication.
1. Organisms. DNA. Replication
I. Title II. Series
574.873282
ISBN 0-19-963216-2

Library of Congress Cataloging in Publication Data
Adams, R. L. P. (Roger Lionel Poulter)
DNA replication/Roger L. P. Adams.
(In focus)
Includes bibliographical references.
Includes index.
1. DNA – Synthesis. I. Title. II. Series: In focus (Oxford, England)
[DNLM: 1. DNA Replication. QH 462.D8 A216d]
QP624.A327 1991 574.87'3282 – dc20 90-15621
ISBN 0-19-963216-2 (pbk.)

Typeset and printed by Information Press Ltd, Oxford, England.

Preface

The replication of DNA, and particularly the factors which limit initiation to a single round of replication per cell division, have fascinated biochemists, cell biologists, and geneticists for generations. Yet understanding has not come smoothly and quickly. Rapid advances are made on one front while another area remains static. Theories are proposed which are later shown to be totally untenable. A book on the subject will reflect the current understanding, but some sections may date very rapidly if the ideas presented are not to be mundane.

Initiation of DNA replication holds the clue to what controls cell division and hence growth and differentiation. At first sight, however, it appears that each and every organism controls DNA replication in a different manner. What is important is to try to draw a common theme from the disparate systems; and this means neglecting a consideration of some plasmid while mentioning another interesting, but perhaps abstruse, observation.

An attempt has been made to give an overview of the subject but this means that many areas and hundreds of references have had to be omitted.

Roger L.P.Adams

Contents

Abbreviations

ARS	autonomous replicating sequence
bp	base pairs
dNTP etc.	any deoxyribonucleoside triphosphate
EBV	Epstein–Barr virus
ECR	extended chromosomal regions
HSR	homogeneously-staining regions
kDa	kiloDalton
NAD^+	nicotinamide adenine dinucleotide
PCNA	proliferating cell nuclear antigen
PCR	polymerase chain reaction
SSB	single-stranded DNA binding protein
Tag	SV40 T-antigen
Ts	temperature sensitive

1

Introduction

1. Semi-conservative replication

DNA consists of two strands of polynucleotides—a duplex. Each strand is a long sequence containing four different deoxyribonucleotides, joined by phosphodiester bonds. The two strands are complementary to each other in that when dAMP occurs in one strand the complementary dTMP occurs in the other strand. Similarly, dCMP occurs opposite dGMP. The structure of the complementary base pairs is shown in *Figure 1.1*.

Replication requires that this structure is copied to produce two duplexes, each identical to the parental duplex. Essentially, this is brought about by the separation of the two parental strands and the building up of two new daughter strands, each complementary to one of the parental strands. This process is known as *semi-conservative* replication as, in each daughter duplex, one of the parental strands is conserved. Two other mechanisms were thought possible: firstly conservative replication, in which one daughter duplex contains the same two DNA strands that were present in the parental duplex, whilst the other daughter duplex is made from entirely new material; or alternatively dispersive replication, in which neither of the parental duplexes remains intact but material from both is dispersed in fragments to the daughter molecules. Evidence that replication occurs by a semi-conservative mechanism was obtained by Meselson and Stahl (1) by first growing *Escherichia coli* cells for several generations in medium containing ^{13}C and ^{15}N, and then following the changes in density of duplex DNA molecules on transfer of the bacteria to normal 'light' medium (*Figure 1.2*).

The precursors used for the synthesis of DNA are the deoxyribonucleoside 5′ triphosphates, which become linked together by phosphodiester bonds with the elimination of pyrophosphate (*Figure 1.3*). This reaction is almost at equilibrium and *in vitro* can readily proceed in the reverse direction given a supply of pyrophosphate. However *in vivo* it is driven in the direction of DNA synthesis

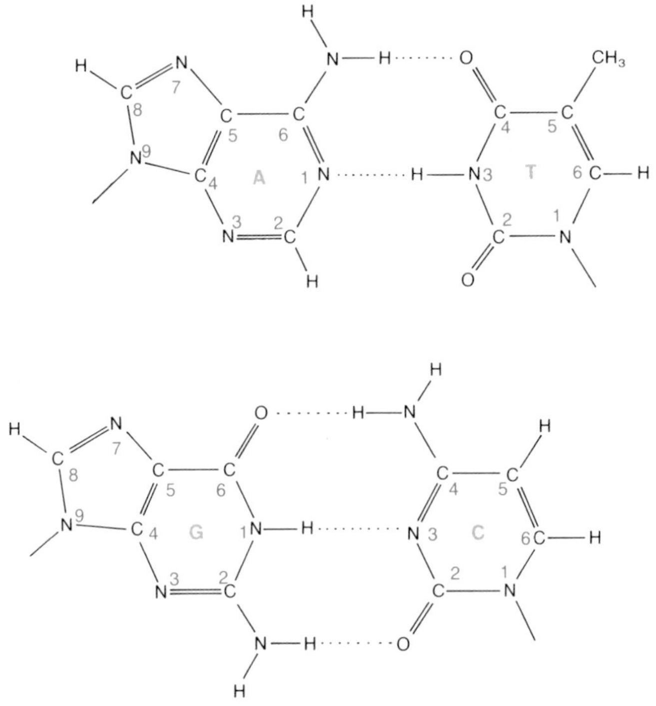

Figure 1.1. Complementary base pairs. The A:T and G:C base pairs are shown, with hydrogen bonds drawn as dotted lines. The bonds from N1 (pyrimidine) and N9 (purine) are the glycosidic bonds to the C1′ of the deoxyribose sugar which is not drawn.

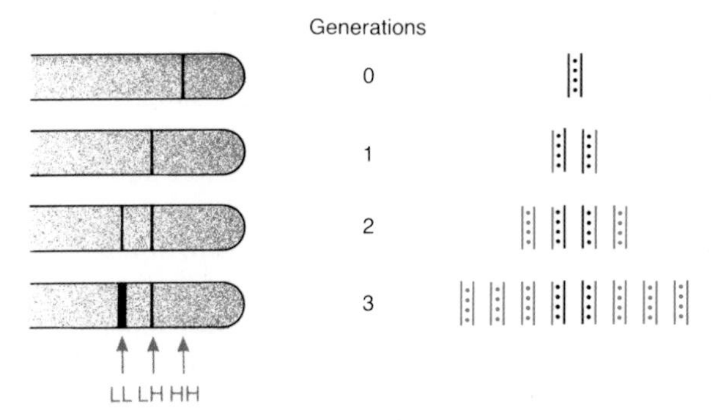

Figure 1.2. Illustration of the Meselson – Stahl experiment. DNA from bacteria, grown in medium containing heavy isotopes, comes to equilibrium well down the tube when centrifuged to equilibrium in CsCl solutions. Both strands of this DNA contain the heavy isotopes at the beginning of the experiment (HH). After growth of the bacteria in medium containing light isotopes for one generation, all the DNA bands at an intermediate position, showing it to be of hybrid density (LH). Continued growth of the bacteria in medium containing light isotopes results in a steadily increasing proportion of the DNA banding at the position of fully light molecules (LL). No material is seen in between these three bands but some of the hybrid density material persists as is expected for semiconservative replication. The interpretation of the results is shown on the right.

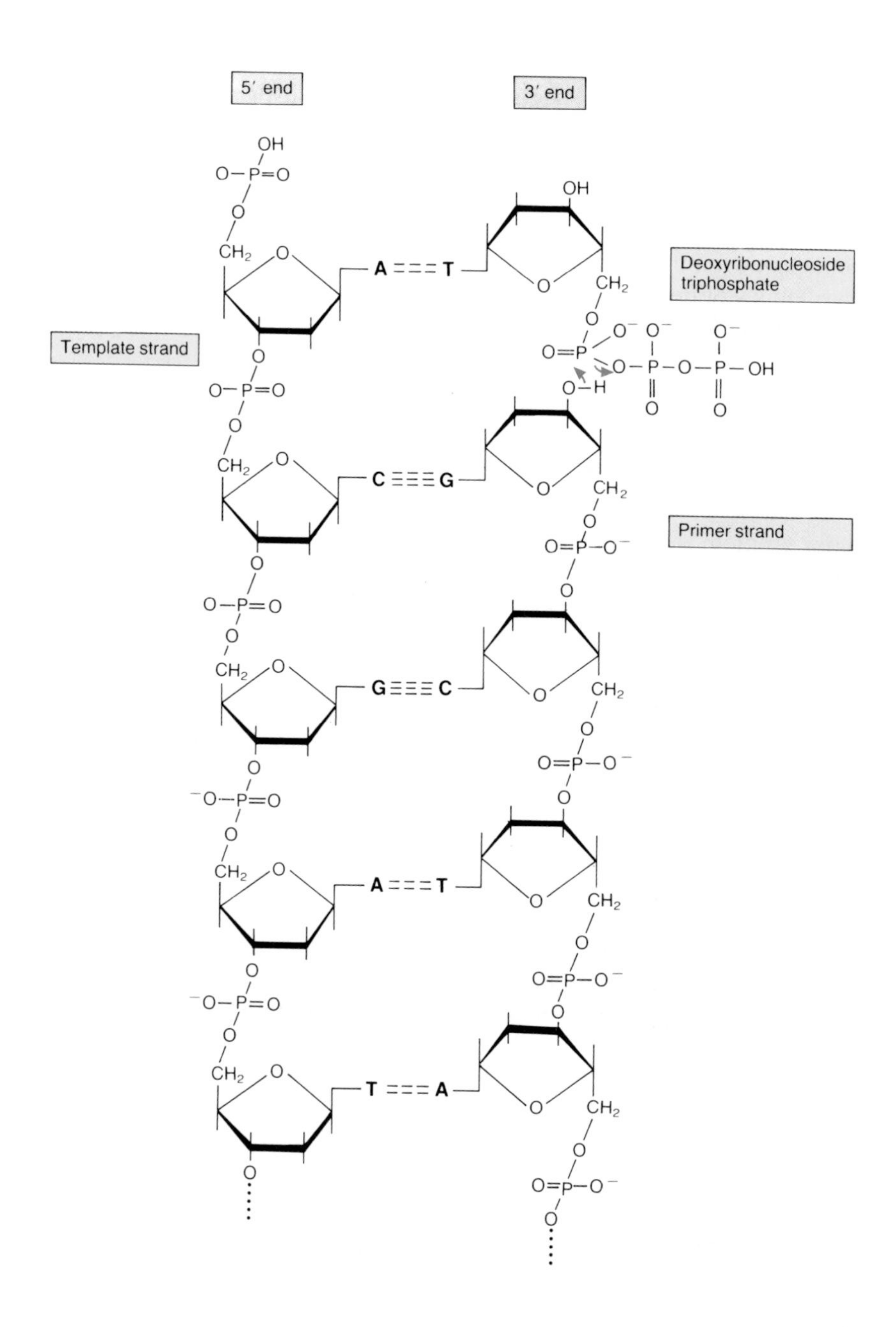

Figure 1.3. The reaction catalysed by DNA polymerase. Note the use of the primer and template strands and the fact that the adjacent ends of the two strands are different.

as a consequence of the removal of the pyrophosphate by hydrolysis to phosphate catalysed by an active and ubiquitous pyrophosphatase.

The enzyme which catalyses the synthesis of DNA is known as *DNA polymerase*. Many DNA polymerases have been characterized and they will be considered in more detail in Chapter 2, Section 7. All DNA polymerases catalyse the joining of a deoxyribonucleotide onto the 3′ OH terminus of an oligonucleotide ***primer*** and the order in which the deoxyribonucleotides are added is dictated by a ***template*** strand (*Figure 1.3*). DNA polymerases cannot initiate synthesis ***de novo*** from mononucleotides but require a primer, which can be an oligoribonucleotide or deoxyribonucleotide. This limitation may be important to prevent unscheduled initiation of replication whenever and wherever a DNA polymerase becomes attached to a template.

2. Problems associated with DNA replication

A major problem is caused by the fact that the two DNA strands in the duplex run in opposite directions. They are said to be ***antiparallel***. This means that the end of one strand is a deoxyribose with a free 3′ OH group whereas the adjacent end of the other strand is a deoxyribose with a free 5′ OH or 5′ phosphate group (*Figure 1.3*).

A second problem is caused by the length of DNA molecules and the instability of single-stranded DNA. This problem is serious for plasmid DNA (about 5000 base pairs long) but is accentuated for bacterial DNA (about 3 million base pairs) and even more so for vertebrate DNA, one molecule of which may be 100 million base pairs in length. This imposes the restriction that the two parental molecules cannot separate along their entire length but can separate only at a restricted region known as the ***replication fork*** (*Figure 1.4*). It is here that the daughter molecules are constructed as the fork progresses along the DNA molecule to create a, so-called, replication bubble. The details of initiation of replication are covered in Chapters 3 and 4, and it is clear that initiation starts at a unique site and usually replication forks move away from the origin in both directions (i.e. replication is bidirectional). Replicating, cyclic molecules, therefore, appear in electron micrographs as θ-shaped structures (see *Figure 3.1b*).

A third problem is brought about by the fact that these very long DNA strands are wound around each other approximately once every 10 base pairs.

3. Continuous and discontinuous synthesis

At the replication fork there are two daughter strands of DNA being synthesized; one complementary to each parental strand. Because of the antiparallel structure of DNA only one of these strands can grow in a 5′ to 3′ direction using the reaction shown in *Figure 1.3*.

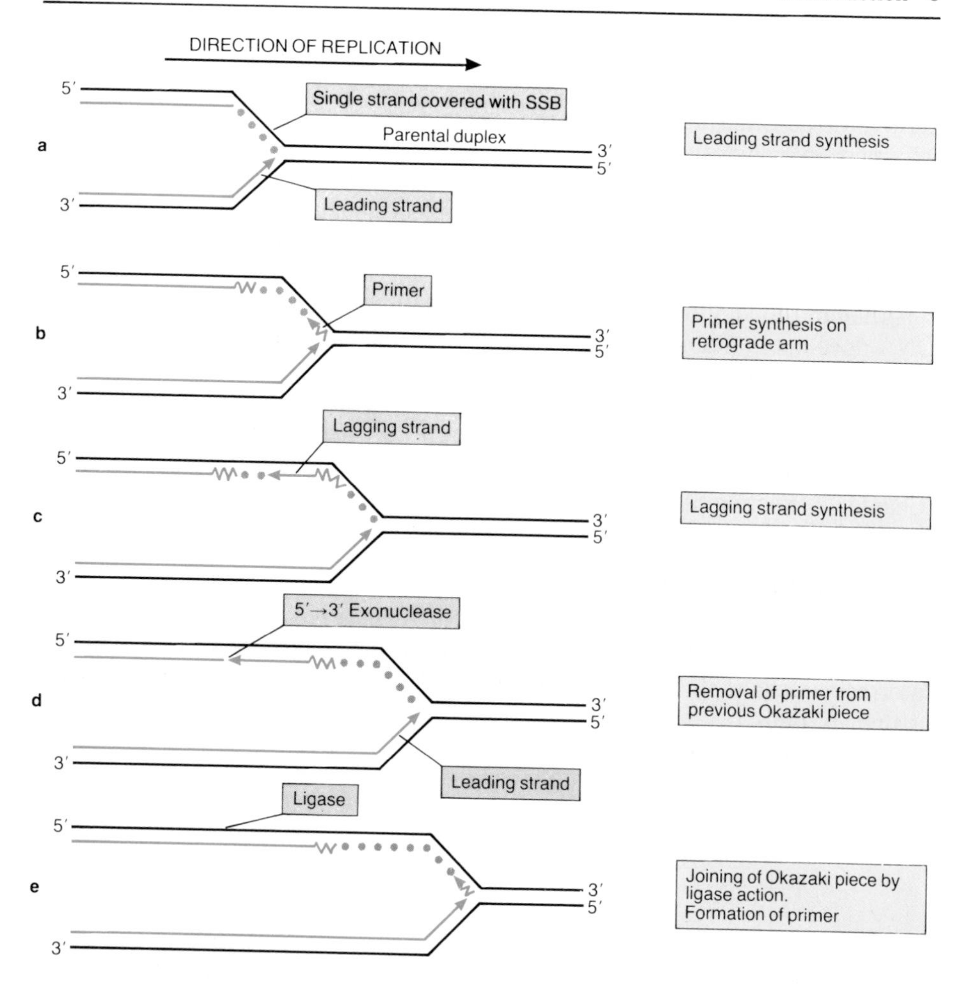

Figure 1.4. The replication fork. The synthesis of the leading strand is accompanied by association of the exposed arm with SSB **a**. A primer is made **b** and the lagging strand is extended to form an Okazaki piece **c**. The previous primer is removed **d** using 5′ to 3′ exonuclease and ligase makes the final join **e**.

At the end of the other growing strand is a 5′ phosphate. This cannot be used as a primer for DNA polymerase and biochemists were perplexed for a long time to know the mechanism of synthesis of this strand. The problem was solved by an ingenious suggestion of Okazaki (2) that the direction of synthesis of this strand was opposite to the direction of fork movement (i.e. retrograde synthesis, *Figure 1.4*). This allows synthesis of the daughter strand in a 5′ to 3′ direction by DNA polymerase, but requires regular initiation of replication of the retrograde strand (*Figure 1.4*). A mechanism must, therefore, be available for repeated generation of a primer on the retrograde side of the fork. This primer is an oligoribonucleotide and it is synthesized by an RNA polymerase specialized for the synthesis of primers, known as a *primase*.

Replication can, therefore, be seen as one daughter strand (the *leading strand*) being synthesized in the direction of replication fork movement. As this strand grows, the duplex is unwound (by the action of a helicase, Chapter 2, Section 3), exposing a region of single-stranded DNA at the other side of the fork (*Figure 1.4*). On this exposed region, primers are made and retrograde synthesis initiated. This second strand is synthesized in small pieces (*Okazaki pieces* or fragments) slightly later than the leading strand, and this strand is therefore known as the *lagging strand*.

Although the lagging strand is made discontinuously using an RNA primer, DNA does not contain multiple nicks, neither does it contain regions of RNA. A further requirement of the Okazaki model is, therefore, that the primers are removed after use and that the Okazaki pieces are joined together. The enzymes required to catalyse these reactions are, a 5′ to 3′ *exonuclease* (to remove the primer), a DNA polymerase (to fill the gap so created) and a *DNA ligase* (Chapter 2, Section 6) to make the final join.

4. Evidence for the Okazaki model

The main features of this model are that short pieces of DNA are formed as intermediates in replication, and that these Okazaki pieces have an RNA primer at their 5′ ends.

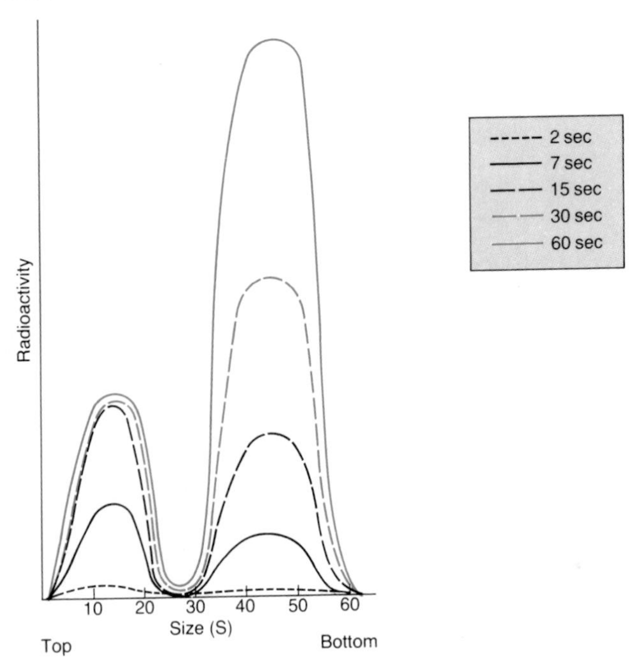

Figure 1.5. Okazaki's experiment (2). Phage T4 infected *E. coli* were labelled for the indicated times with tritiated thymidine and the isolated DNA denatured with alkali and sedimented through a sucrose gradient.

Okazaki obtained evidence for the former by labelling bacteriophage (phage) T4 infected *E. coli* with tritiated thymidine for various times (2). To isolate the Okazaki fragments it was necessary to release them from the high molecular weight template DNA and this was achieved by alkaline denaturation. The small intermediates were then separated from the larger DNA by sedimentation on sucrose gradients. After short labelling times the major part of the radioactivity was found in pieces of DNA of about 1500 nucleotides long. With increasing labelling times, the amount of radioactivity in these small pieces quickly levelled off while the radioactivity in the bulk of the DNA continued to increase, providing evidence that small lengths of DNA are intermediates in DNA replication (*Figure 1.5*). Similar experiments have been done in eukaryotes, but here the Okazaki pieces were found to be only about 200 nucleotides long.

Because of technical problems, the evidence for an RNA primer was more difficult to obtain. Evidence had been obtained that the majority of Okazaki pieces

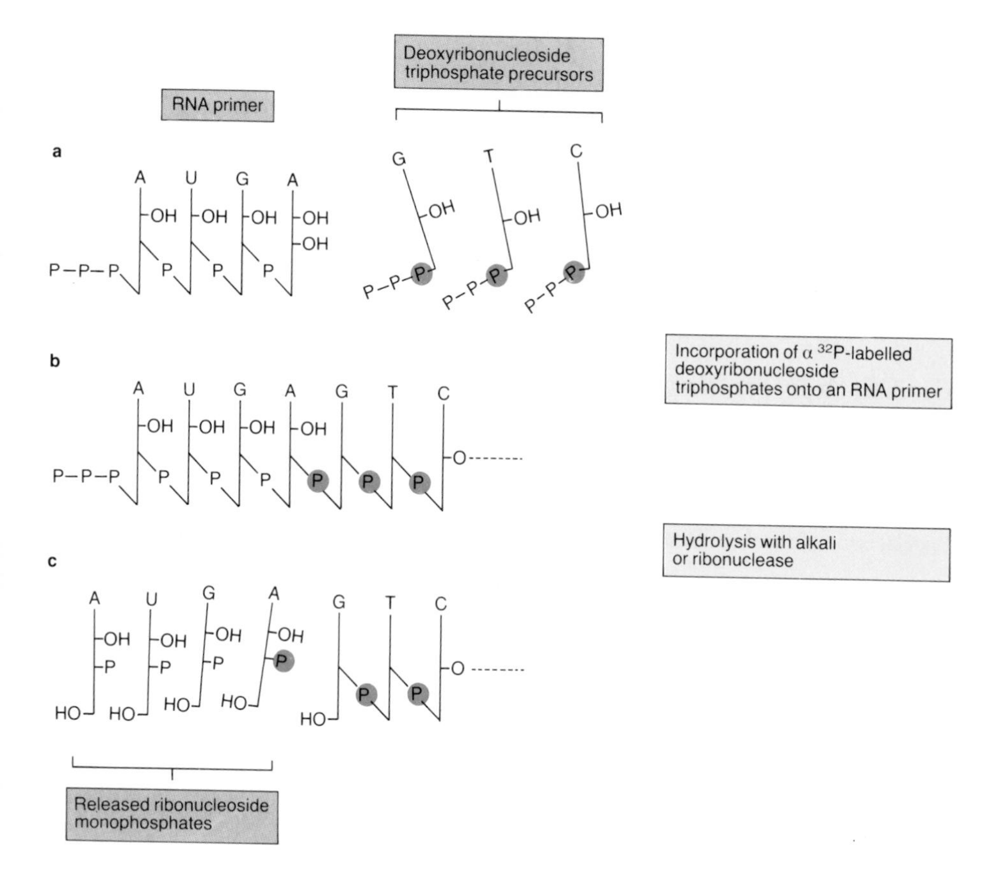

Figure 1.6. Nearest neighbour analysis of the primer junction. Incorporation of α^{32}P labelled deoxyribonucleotides is followed by alkaline hydrolysis of the RNA primer leaving the 3′ nucleotide of the primer radioactive.

had something (presumably RNA) at their 5′ ends which was labile to alkali and to ribonuclease. However, it took several years to devise methods for DNA denaturation which do not involve alkali (which degrades RNA) and which do not result in non-specific association of RNA with single-stranded DNA. Okazaki pieces isolated in the presence of formamide were shown to have ribonucleotides attached to their 5′ ends; and removal of the DNA component showed the primer to be only 2–4 nucleotides in length in *E. coli* but about 10 nucleotides long in vertebrate cells (3,4).

A variety of tricks were employed to obtain DNA replication in the presence of [α-^{32}P] *deoxyribo*nucleoside triphosphates (dNTPs), and this led to evidence for the transfer of the alpha position phosphate group to a *ribo*nucleotide. This could only occur if there is a covalent link between the RNA primer and the DNA (*Figure 1.6*). By using each labelled dNTP in turn, and by analysing the nature of the radioactive ribonucleotide generated it was concluded that there was no sequence specificity at the RNA/DNA junction (5).

5. Conclusion

This chapter has provided the background information on DNA structure and semiconservative replication. From the evidence presented it is clear that replication occurs at a replication fork, with the leading strand growing continuously and the lagging strand being made as short, RNA-primed, Okazaki pieces which are subsequently joined together.

6. Further reading

Adams,R.L.P., Knowler,J.T. and Leader,D.P. (1986) *The Biochemistry of the Nucleic Acids* (10th edn), Chapter 6. Chapman & Hall Ltd.

Kornberg,A. (1980) *DNA Replication*. W.H.Freeman and Co.

Special issue of *Biochim. Biophys. Acta* (1988) *Enzymology of DNA Replication*, **951, No. 2/3**.

7. References

1. Meselson,M. and Stahl,F.W. (1958) *Proc. Natl. Acad. Sci. USA*, **66**, 671.
2. Okazaki,R., Okazaki,T., Sakabe,K., Sugimoto,K., Kainuma,R., Sugino,A. and Iwatsuki,N. (1968) *Cold Spring Harbor Symp. Quant. Biol.*, **33**, 129.
3. Ogawa,T., Hirose,S., Okazaki,T. and Okazaki,R. (1977) *J. Mol. Biol.*, **112**, 121.
4. Reichard,P., Eliasson,R. and Soderman,G. (1974) *Proc. Natl. Acad. Sci. USA*, **71**, 4901.
5. Pigiet,V., Eliasson,R. and Reichard,P. (1974) *J. Mol. Biol.*, **84**, 197.

2

Enzymes of replication

1. Introduction

This introduction will describe the need for a number of proteins which are required at the replication fork. Subsequent sections will then describe the proteins in more detail.

The names of some of the proteins involved in replication (e.g. DnaB) can be confusing, but their names reflect the methods involved in their discovery (Section 8). For example, a number of *E. coli* mutants were isolated that were defective in DNA replication: one of these was called *dna*B. Later, the protein which was defective in the mutant was isolated and called DnaB protein. Only later still was the function of this protein established, when it became known as DnaB helicase.

From a consideration of Chapter 1 it is clear that DNA polymerase, primase, 5′ to 3′ exonuclease and DNA ligase are all required to replicate DNA. In addition, *helicases* are required to unwind the DNA ahead of the point of replication and single-stranded-DNA-binding proteins (*SSBs*) are required to bind to and stabilize the single-stranded regions of DNA arising on the retrograde or lagging arm at the fork.

In a cyclic, covalently-closed DNA molecule, it is easy to imagine the complications which will occur as the two, linked strands of DNA are unwound. Unwinding in one region must lead to overwinding in another region unless one strand is broken and allowed to rotate with respect to the other. Such a swivel can be produced by the action of a *topoisomerase*; an enzyme which alters the topological linking number of DNA. Rotation of the DNA duplex is also constrained in linear molecules and the need for topoisomerases is universal. This is true not only during replication but also during transcription.

DNA synthesis is a very rapid process. In *E. coli* the replication fork moves at 800 base pairs per second (Chapter 3, Section 2.2). The organization required to achieve this complex reaction does not allow time for DNA polymerase to work in the classic mode of an enzyme. That is to say, the polymerase cannot leave the DNA after addition of each nucleotide (a *distributive* mode of action). Rather, the enzyme must always remain associated with the terminal 3′ OH group. This means that one molecule of polymerase must process along the DNA template adding all the nucleotides to the leading strand. To achieve this *processive* mode of action requires the presence of several supplementary proteins. The problem is worse in the case of the lagging strand which obviously must achieve the same overall rate of synthesis as the leading strand. In this case, not only must the polymerase be processive, but it must also cycle rapidly from the end of one Okazaki piece to the beginning of the next. The details of how this is achieved are only now being worked out, but the current model is presented in Section 7.2.

The fidelity of replication is very high. Overall, only one mistake is made for each 10^{12} nucleotides copied. This high level of fidelity is achieved in several stages, the first of which is non-enzymic. Thus, the stability of the possible base pairs differs and the GC base pair is much more stable than a GT base pair, and AT is more stable than AC. Although it is clear that some of the less stable pairings do occur under certain circumstances, the overall stability of the Watson–Crick pairings alone would allow replication to occur with only one error per thousand nucleotides incorporated.

The other mechanisms which ensure the high fidelity of replication are largely a consequence of the properties of the DNA polymerase or closely associated proteins (Section 7.1). The exceptions to this are mismatch repair mechanisms which come into play shortly after replication is complete and which serve to increase fidelity from one error in 10^9 bases replicated to one in 10^{12} bases replicated. Section 7.4 describes some of the methods used to measure fidelity and gives some conclusions.

2. Single-stranded DNA binding proteins (SSBs)

The affinity of these proteins for single-stranded DNA is several orders of magnitude greater than for duplex DNA. The SSB coded for by gene 32 of phage T4 was the first to be studied and this is now used to help visualize single-stranded DNA in the electron microscope. It binds cooperatively to DNA, each protein molecule covering about 4 nucleotides, and autoregulation of synthesis ensures that there are just enough molecules present in a cell to cover the 1500 bases maximally exposed at each replication fork (*Figure 1.4*) (1). Whatever its origin, single-stranded DNA will be covered by SSB; for example, when *E. coli* is infected with a single-stranded DNA phage, such as M13 or ϕX174, the incoming DNA is immediately complexed with SSB.

3. Helicase

Many different helicases have been characterized and at least six different enzymes are present in each *E. coli* cell. These are proteins which also bind to single-stranded DNA, but not usually in a cooperative manner. Rather, after binding, a molecule of helicase will travel along the DNA driven by the energy provided by the hydrolysis of ATP (2). Travel is unidirectional, and there are some helicases which travel in a 3′ to 5′ direction (e.g. Rep, helicase II, and n′) whereas others (e.g. helicase I and DnaB) travel in a 5′ to 3′ direction. When the helicase encounters a region of duplex DNA unwinding occurs (*Figure 2.1*). It is important to realize that a helicase does not bring about a change in the linking number of a DNA molecule, and that extensive helicase action will also normally require an SSB to bind to the single-stranded DNA generated, and a topoisomerase to act as a swivel.

It appears that DnaB helicase cannot normally bind directly to DNA covered in SSB. At least two alternative mechanisms are available to obtain binding. One, which requires the DnaA protein, is involved in initiation of replication (see Chapter 3). The other mechanism is important in the initiation of Okazaki pieces on the lagging side of the replication fork. It involves prior binding of the n′ protein which itself binds only to a recognition sequence which occurs occasionally in DNA. This n′ binding sequence involves a hairpin loop which is present in a region of otherwise single-stranded DNA. Although the sequence of bases in the stem of the loop is not critical, the sequence AAGCGG is required in the single-stranded region at the top of the loop. Once located on the DNA, n′ protein binds to a series of other proteins (n, n″, i, DnaC) in an ATP-requiring reaction. This complex is now able to bind to a hexamer of DnaB protein molecules and the DnaC protein is displaced. Only now is the DnaB complex (or ***preprimosome***) able to translocate along the single-stranded DNA to which it is bound (see *Figure 2.3*) (3). These difficulties placed in the way of binding of the DnaB helicase must be required to ensure that unscheduled unwinding of the DNA does not occur. Binding is hereby restricted to the period of initiation of replication and the specific times and places where replicating forks are growing (4).

The preprimosome contains two helicases (n′ and DnaB) which move in opposite directions along the bound strand. They are both associated with the same strand and are tightly bound together. The result of their joint action may be to generate a loop of DNA on the lagging strand, just behind the replication fork (see Section 7.2).

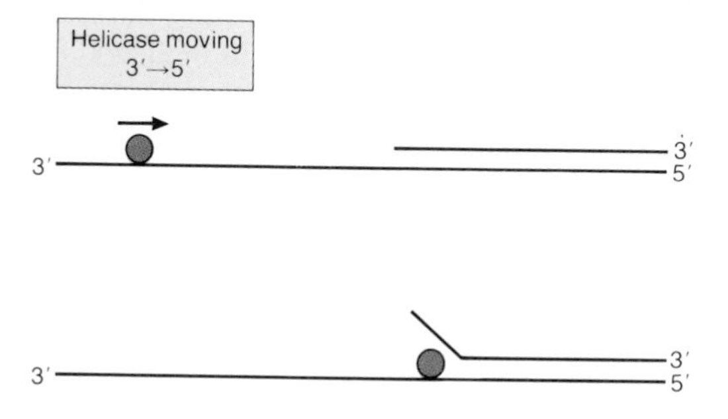

Figure 2.1. Helicase action.

Helicase activity can be assayed by measuring the displacement of short (30–40 mer), ^{32}P-labelled oligonucleotides from complementary single-stranded circles of M13 DNA. Displacement can be quantitated by counting the released oligonucleotide after electrophoresis of the deproteinized mixture (5).

4. Topoisomerase

A topoisomerase changes the linking number of a DNA duplex, that is, it introduces (or removes) Watson–Crick turns. However, it achieves this without producing a permanent break in either strand of the duplex DNA. The original model proposed that the mechanism of action involved nicking one of the DNA strands, rotating a free end around the intact strand, and religating to seal the nick; all this occurring without releasing from the enzyme either of the ends at the nick. It is now realized that, rather than a rotation being involved, the same effect can be achieved if the intact strand is passed through the nick which is then resealed.

There are two major types of topoisomerase. Type I catalyses the reaction described above and changes the linking number one at a time. A type II topoisomerase changes the linking number two units at a time and acts by breaking both strands of a duplex molecule and passing another duplex through the gap before it is resealed.

Type I topoisomerases do not require a source of energy and serve to relax DNA which has a deficiency of Watson–Crick turns (i.e. negatively supercoiled DNA). They can thus work in conjunction with a helicase to unwind the duplex DNA at the replication fork. They also play a role in transcription.

Type II topoisomerases require ATP. As with type I topoisomerases, they can relax negatively supercoiled DNA but, in addition, they are used to separate two *catenated* (interlinked) DNA duplexes such as those which arise on completion of replication of cyclic DNA molecules and at meiosis (see Chapter 5, Section 2.1). Some type II topoisomerases are able to act in reverse (e.g. the DNA gyrase from *E. coli* can not only remove negative superhelical turns from DNA but it can also introduce such turns) and a reverse gyrase from archaebacteria can introduce positive superhelical turns into a closed circular duplex DNA molecule.

DNA topoisomerases can be assayed by measuring the steady rate of removal of superhelical turns from a supercoiled plasmid molecule or DNA from SV40 (*Figure 2.2*) (6). A type I topoisomerase removes superhelical turns one at a time, whereas a type II enzyme removes two superhelical turns at a time. In contrast, an endonuclease would completely relax the DNA in a single step and would produce a nicked duplex.

Topoisomerases act by breaking a phosphodiester bond and transferring the 3′ (type I) or the 5′ (type II) terminal phosphate group to a tyrosine of the protein. Such complexes containing protein covalently linked to DNA can be isolated by adding SDS to the enzymic reaction. Certain inhibitors of topoisomerases (e.g. camptothecin, teniposide) exert their effect by preventing the subsequent

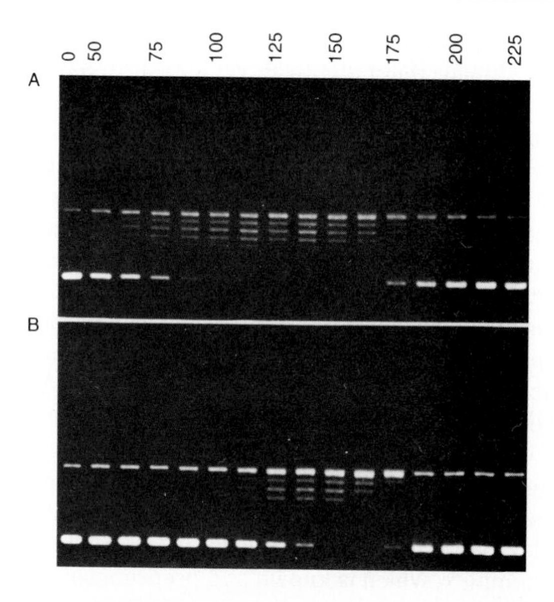

Figure 2.2. Topoisomerase assay. Yeast topoisomerase I (A) and II (B) were incubated with plasmid DNA in the presence of 0.2 mM ATP and increasing concentrations of KCl (the concentration present in alternate incubations is given in mM across the top). The reaction mixture was then fractionated on an agarose gel to separate the topoisomers. From (6) with kind permission of the authors and publisher.

step of strand passage and hence lead to the accumulation of the DNA:protein intermediate (7, 8). Type I enzymes are monomers but type II topoisomerases are dimeric enzymes which cut the DNA duplex with a 4-base stagger. From a study of the sequence of DNA around the protein attachment site it has become clear that topoisomerases do not act at random. The *Tetrahymena* topoisomerase I cleaves at the sequence A(A/G)ACTT*AGA(G/A)AAA(T/A)(T/A)(T/A) but, although the vertebrate enzyme prefers this hexadecamer, it will also cleave at other sequences and, in SV40 chromatin from camptothecin-treated cells, breaks are found at or near the replicating forks (9). Yeast topoisomerase II acts predominantly at oligo(A) tracts, although the *Drosophila* and vertebrate enzymes show different specificity and will act at sequences of alternating purines and pyrimidines (10).

5. DNA primase

The *E. coli* DnaG primase was the first to be described, but the eukaryotic enzyme and that of phage T7 have a number of different features.

The phage T7 primase (the product of T7 gene 4) is present as a dimer which also has DNA helicase activity and which interacts specifically with the phage

T7 DNA polymerase (see Chapter 3, Section 3.4.1). The two gene 4 protein subunits are very similar, if not identical, yet appear to have different functions (11). Unlike the *E. coli* and eukaryotic enzymes the phage T7 helicase/primase makes a primer of specific sequence pppACC(C/A). Thus on replication of phage T7, primers are made only when the complementary sequence occurs on the template strand on the lagging side of the fork. It is along this strand that the phage T7 helicase/primase is moving, so that when a primase recognition site is encountered one of the subunits dissociates to synthesize the primer and then presumably rejoins the helicase subunit (see *Figure 3.10*).

E. coli DnaG primase can use either ribo- or deoxyribonucleoside triphosphates to synthesize a primer but, when both are present, it will make a ribonucleotide primer 3–5 nucleotides long before handing over to DNA polymerase (12).

Only in a few exceptional circumstances can *E. coli* primase bind to DNA. Thus it can bind to a complex, hairpin-loop structure present in the single-stranded DNA of the phages G4 and ϕK, and this is important for initiating the synthesis of the replicative form of these phages (see Chapter 3, Section 3). Normally, it binds to a complex containing the DnaB helicase (see Section 3.3). As described previously, this complex, which is known as a preprimosome, translocates along single-stranded DNA in a 5′ to 3′ direction (3), but may also form looped structures. At certain, non-random sites (whose sequence requirements have not yet been determined) the DNA that is looped around the complex can bind DnaG primase to form a ***primosome*** (*Figure 2.3*). A primer is initiated and, as the primase

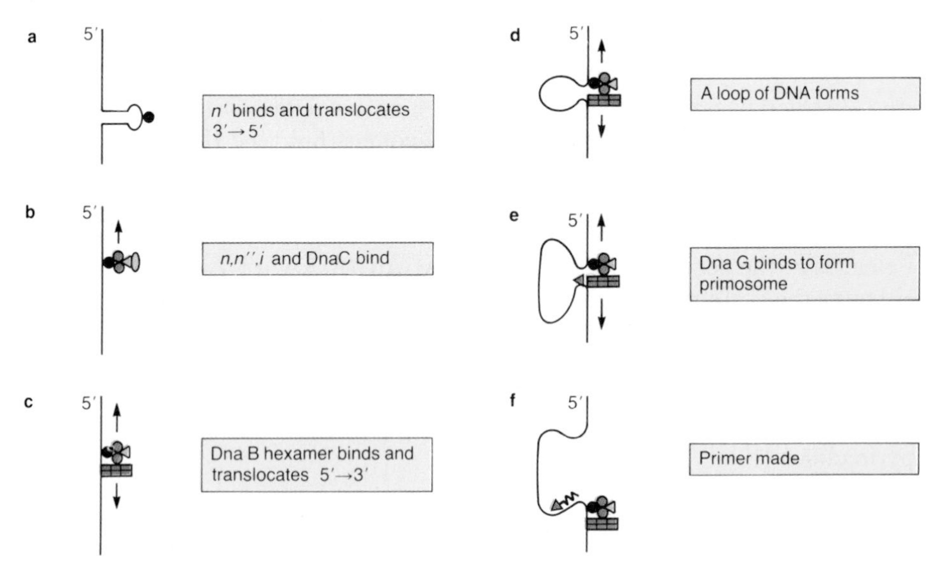

Figure 2.3. Primosome formation. n′ binds to its recognition sequence at the top of a hairpin loop **a** and may translocate in a 3′ to 5′ direction. Binding of n, n″, i, and DnaC **b** is required to get binding of a DnaB hexamer to form the preprimosome **c** which might translocate in both directions along the single-stranded DNA forming a looped structure **d**. At this point DnaG primase binds and a primer is made **e,f**. The preprimosome continues to translocate along the DNA in the overall direction 5′ to 3′.

moves away around the loop in a 3′ to 5′ direction along the bound, template strand, the released preprimosome is again free to translocate in the 5′ to 3′ direction. Once again, as with the T7 primase/helicase, there is a close association of primase and helicase which can travel along the lagging side of the fork, unwinding the DNA and initiating primers when suitable recognition sites are encountered.

The eukaryotic primase appears simpler, but occurs in a complex with DNA polymerase α (see Section 7.3) (13). It initiates primers largely with dATP but, although it works most efficiently at sites where the sequence 3′CCC occurs 10 bases downstream (14), it may be able to bind to almost any region of single-stranded DNA. As eukaryotic Okazaki pieces are approximately the size of a nucleosome repeat it is likely that chromatin structure places some limitations on priming activity in eukaryotes (see Chapter 4).

6. DNA ligase

This enzyme is able to join nicks in one strand of a double-stranded DNA molecule. The enzymes found in most species act fundamentally in the same way but, at high concentrations, the phage T4 DNA ligase is also able to join two blunt-ended DNA molecules.

The nicks must be bounded by 3′ OH and 5′ phosphate groups. The reaction usually requires ATP (the *E. coli* DNA ligase uses NAD^+ instead), the AMP part of which becomes covalently linked to the enzyme before being transferred to the 5′ phosphate at the nick. A pyrophosphate bond is thereby created. The 3′ OH group now makes a nucleophilic attack on this, displacing the AMP and joining the two parts of the DNA molecule together.

7. DNA polymerases

Here lies the key to DNA replication as these enzymes catalyse the formation of the phosphodiester bonds required for synthesis of DNA. DNA polymerases are also involved in DNA repair and recombination and all cells have a multiplicity of enzymes each specialized for use in a different circumstance. They mostly act in complexes along with other enzymes as indicated in Section 1. Many different enzymes have been characterized but, in order to avoid confusion, just four, which illustrate the salient features of DNA polymerases, will be considered in this section. Others, for example some virally-coded enzymes, will be referred to in Chapter 3 but, for a more comprehensive treatment, consult reference 15.

7.1 E. coli *DNA polymerase I*

E. coli DNA polymerase I, otherwse known as the Kornberg enzyme after the man who first detected and purified it, is only occasionally involved in synthesis of long stretches of DNA. Rather it is required to remove primers from Okazaki pieces and to fill in the gaps so produced. To do this it requires a 5′ to 3′ exonuclease activity and an ability to attach to DNA at a nick or very small gap (*Figure 2.4*). It does not work in association with the primosome but it can use a longer RNA primer as described in Chapter 3, Section 5.2.

The enzyme is a single, large polypeptide but it can be readily cleaved into two fragments. The larger, Klenow fragment retains the polymerase activity and 3′ to 5′ exonuclease activity, but not the 5′ to 3′ exonuclease activity.

The Klenow fragment has been crystallized and the DNA shown to lie in a deep groove almost buried by encircling arms (16). There is a region to recognize the template strand, a 3′ OH binding site, and a site to which incoming triphosphates bind. The base bound at the template site is thought to modify the enzyme such that the K_m for the correct triphosphate is lowered by perhaps three orders of magnitude. This, together with the ΔG effect described in Section 1, decreases the chance of a misincorporation to about one in a million.

Nevertheless, mistakes do occur and result in an incorrect nucleotide being present in the primer site. This is a poor substrate for subsequent nucleotide addition. The affinity of binding to the primer site is reduced at the same time as the affinity of binding to an adjacent site is enhanced. The terminal nucleotide therefore moves to the second site where a 3′ to 5′ exonuclease begins to digest the primer strand (17). Having removed the terminal, mismatched base the situation is reversed and the correctly paired terminus now has enhanced affinity for the polymerase site. The combination of polymerase and 3′ to 5′ exonuclease thus acts to *proofread* the DNA as it is synthesized. Proofreading enhances the fidelity of replication by approximately another three orders of magnitude.

$PolA^-$ mutants are viable showing that *E. coli* DNA polymerase I is not essential for bacterial survival. It is clear, therefore, that the functions of DNA polymerase I can also be performed by DNA polymerase III with the help of the less-well studied DNA polymerase II. However, the lack of an efficient 5′

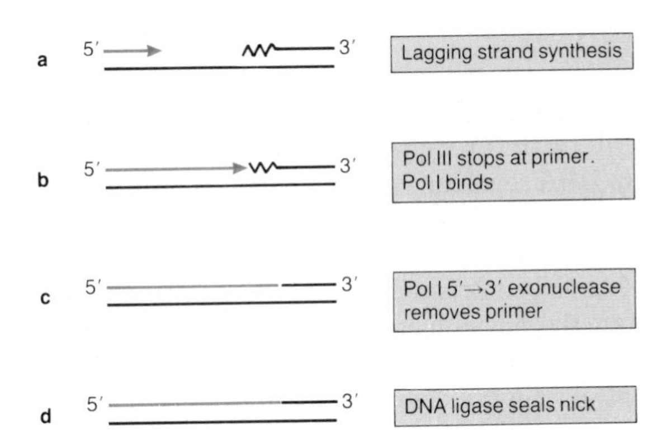

Figure 2.4. 5′–3′ exonuclease and primer removal. In lagging strand synthesis, DNA polymerase III is displaced from the growing Okazaki piece when it encounters the 5′ end of the previous primer **b**. The primer is removed by the 5′ to 3′ exonuclease action of DNA polymerase I which also fills in the gap **c** for ligase to seal **d**.

to 3′ exonuclease in the mutant results in a slower joining of the Okazaki pieces (*Figure 2.4*) and an increased susceptibility of PolA$^-$ cells to agents which damage DNA.

7.2 E. coli *DNA polymerase III*

DNA polymerase III is the enzyme responsible for the bulk of DNA replication in *E. coli*. It is the α subunit of a multisubunit complex known as holopolymerase III. Three subunits (α, ϵ, and θ) form the core polymerase which also has 3′ to 5′ exonuclease activity residing in the ϵ subunit.

This proofreading exonuclease is modified in the RecA protein mediated SOS response which follows damage to DNA (18). The resulting 'error prone' synthesis is important to enable the polymerase to synthesize a daughter strand where the template contains a damaged, modified, or absent base.

The DNA holopolymerase III is seen by Kornberg as a V-shaped enzyme, where the two arms have different compositions and functions (19). Each arm contains a copy of the core enzyme joined by the θ subunits. One arm (the left hand arm in *Figure 2.5*) contains two copies of the τ subunit which increases processivity to such an extent that one molecule of enzyme can extend a primer by hundreds of thousands of nucleotides without ever being released into solution. This is exactly what is required for leading strand synthesis.

The right hand arm of the holopolymerase III contains two copies of the γ complex, each consisting of five proteins, γ, δ, δ', χ, and Ψ. The function of this

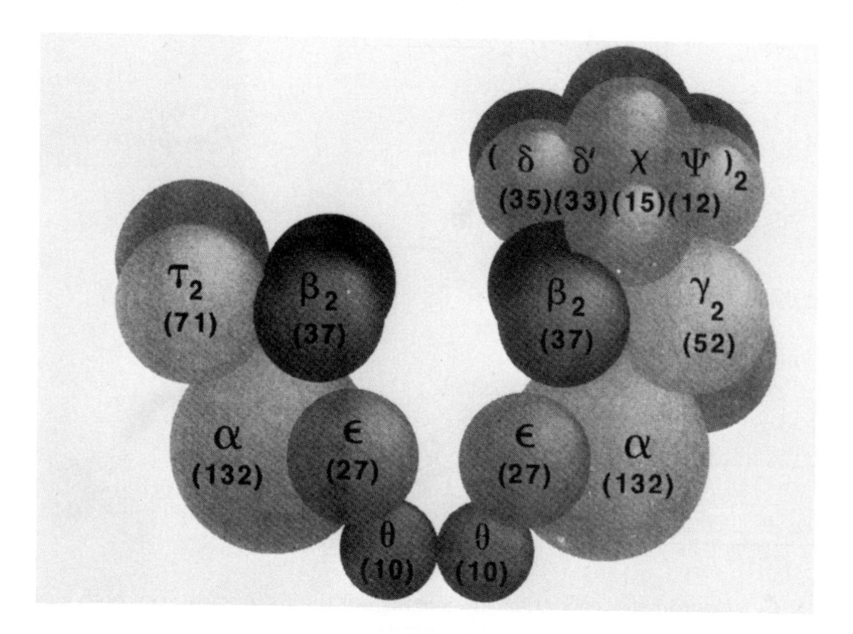

Figure 2.5. Polymerase III holoenzyme. A schematic representation of the holopolymerase III complex. From (22) with kind permission of the authors and publishers.

complex is to facilitate transfer of the enzyme from the end of one Okazaki piece to the primer at the start of the next Okazaki piece. It is not known how this is achieved, but it is required on the lagging side of the fork (20).

In addition, each arm of the V-shaped molecule is loosely associated with two molecules of the β subunit, otherwise known as copolymerase III. This is a helicase which travels in a 3′ to 5′ direction along the bound strand and which could serve to unwind any local duplex regions which might form by foldback of the single-stranded DNA at the replicating fork.

If the left-hand arm of the holoenzyme is catalysing leading strand synthesis and the right-hand arm lagging strand synthesis, how can the enzyme be in two places at once—places separated by up to 1500 nucleotides? The answer lies in the structure of the DNA at the replicating fork (21). As shown in *Figure 2.6*, by bending the single-stranded DNA on the lagging side of the fork into a loop, two objectives are achieved. Firstly, the site of replication of leading and lagging strands is brought into close proximity and secondly, on completion of one Okazaki piece the holopolymerase ends up at a suitable site for initiation of the next.

Although there are only 10 molecules of polymerase III per *E. coli* cell such a holopolymerase III complex, along with the primosome and possibly helicase II or rep, is capable of catalysing replication at the fork. The products of this

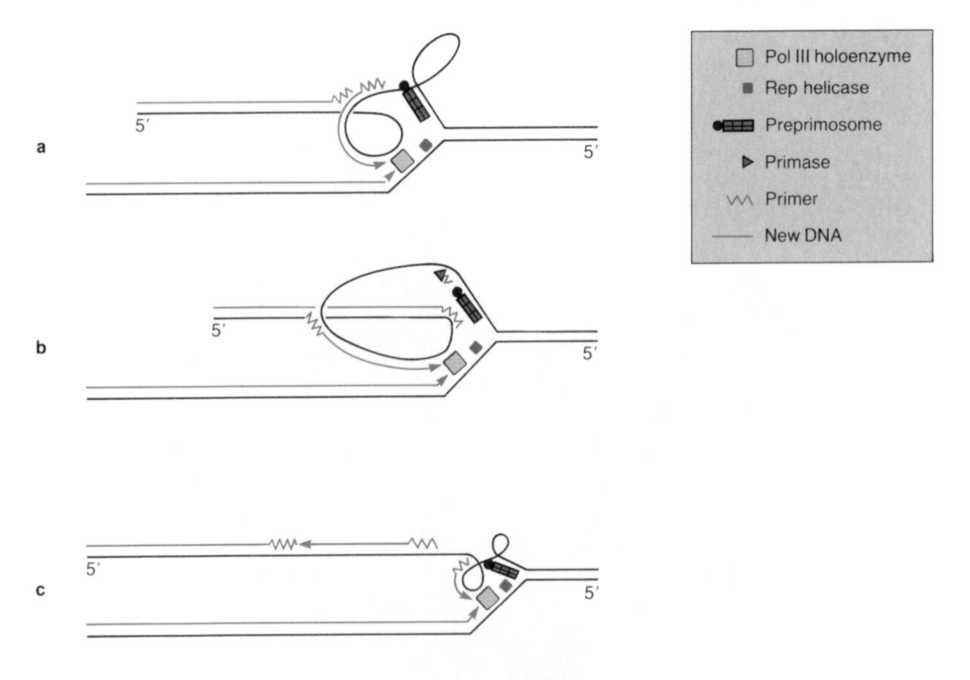

Figure 2.6. Coordinate replication. By looping out the retrograde arm, both leading and lagging strand synthesis can be accommodated on a single enzyme complex **a**. As an Okazaki piece nears completion a new primer is made near the fork **b**. The lagging strand readily transfers to the new primer **c**.

action are DNA duplexes in which the leading strand is continuous but the lagging strand is made of RNA-primed Okazaki pieces. The intervention of DNA polymerase I is required to remove the primers and fill in the gaps so created. DNA ligase will then synthesize the final phosphodiester bond.

7.3 Eukaryotic DNA polymerases

Eukaryotic cells contain several DNA polymerases of which DNA polymerase α and DNA polymerase δ are believed to act at the replicating fork in a complex similar to the *E. coli* holopolymerase III (22, 23). In addition, cell organelles have their own DNA polymerases and infecting viruses often code for a virus-specific DNA polymerase.

Polymerase α is a 180 kDa protein isolated along with three other proteins. Two of these (of 50 and 60 kDa) constitute a primase. This complex is thus capable of initiating and synthesizing Okazaki pieces on the lagging side of the fork.

Polymerase δ is similar in many ways to polymerase α, being susceptible to the same inhibitors, though at different concentrations. It has a more overt proofreading exonuclease than does polymerase α and, in S-phase cells, is found associated with a cyclin known as 'proliferating cell nuclear antigen' or PCNA. This increases the processivity of DNA synthesis and this complex is believed to be involved in leading strand synthesis.

The combination of the two polymerases can therefore act (with other proteins—ref. 24) at the replicating fork to synthesize both daughter strands of DNA. As the rate of fork movement is much slower in eukaryotes (only 50 nucleotides/sec) and the amount of polymerase is greater than in prokaryotes there is less need for complete processivity. On the other hand, multiple replicons are active (see Chapter 4, Section 1) and replication usually takes place on a chromosome matrix site (see Chapter 4, Section 7) resulting in a far more complex situation than is present in prokaryotes.

7.4 Fidelity assays

One of the most popular assays of polymerase fidelity involves the use of an amber mutant in the region of phage ϕX174 containing genes A and B (25) (*Figure 2.7*). The amber codon is lost by mispairing at any one of the three sites, and three different mispairings are possible at each site. Analysis of plaque morphology and activity of genes A and B indicates the degree and type of mispairing which has occurred. This assay is limited in that all the mispairings being studied are occurring in the same sequence and so nearest neighbour effects cannot be studied. An alternative assay (26) involves measuring the extent of substitution of an incorrect nucleotide in a polymerase reaction containing three dNTPs and one dideoxyNTP.

Initially it was thought that eukaryotic DNA polymerases lacked proofreading exonucleases and it was unclear how they managed to replicate DNA as faithfully as did prokaryotic enzymes. The results even seemed to cast doubts on the importance of proofreading as a mechanism of maintaining fidelity. With the

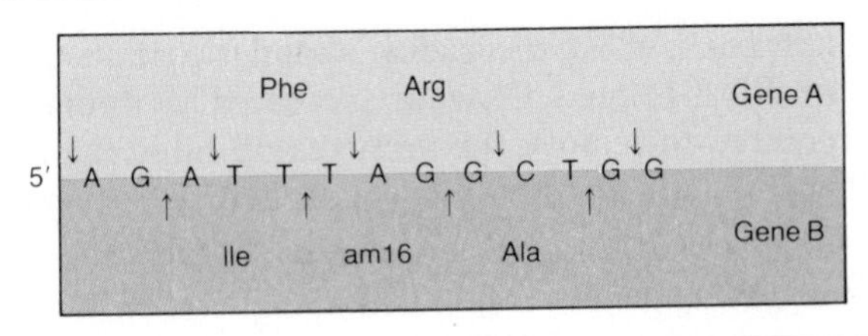

MISPAIRING	T–G	T–T	T–C (wt)	CORRECT T–A
Gene A	Phe•Arg	Leu•Arg	Leu•Arg	Phe•Arg
Gene B	Glu	Lys	Gln	amber

MISPAIRING	A–G	A–A	A–C	CORRECT A–T
Gene A	Phe•Arg	Phe•Trp	Phe•Gly	Phe•Arg
Gene B	Ser	Leu	Trp	amber

MISPAIRING	G–G	G–T	G–A	CORRECT G–C
Gene A	Phe•Thr	Phe•Lys	Phe•Met	Phe•Arg
Gene B	Tyr	ochre	Tyr	amber

Figure 2.7. Fidelity assay. This assay utilizes an amber mutation in a region of ϕX174 DNA which, in the wild type, codes, in different reading frames, for the products of genes A and B. The table shows the consequences of mispairing (and correct pairing) within the amber codon on the amino acid sequences of the two proteins. Eight of the nine possible mispairings will remove the stop codon and allow growth which can be assessed by plaque assay. The plaque morphology differs between the different mutants allowing analysis of individual mispairings.

discovery of the cryptic 3′ to 5′ exonuclease in most, if not all, polymerases, this confusion was resolved and it is clear that the supply of a balanced pool of the four dNTPs is important to maintain maximum fidelity.

8. Methods used in detection and isolation of replication proteins

In order to replicate DNA there is clearly a need for a DNA polymerase but the need for many of the proteins now known to be involved was not predicted. Indeed, even now, the precise functions of some of the proteins required is not known.

Some idea of the complexity of the system came from the isolation of a number of conditional lethal (temperature sensitive) mutants of *E. coli* which were defective in DNA replication. Such mutants indicate the involvement of proteins coded for by the *dna*B, *dna*C, *dna*G, and several other genes. These *dna* mutants do not make DNA *in vivo* at the restrictive temperature; nor do lysates prepared from them synthesize DNA *in vitro*. However, a defective lysate from a *dna* mutant cell can be complemented with a small amount of an extract from the wild type cell. This provides an assay for the protein which is absent from the mutant lysate. Such an assay is called an *in vitro* complementation assay and requires no prior knowledge of the nature or function of the protein, other than that it is required for DNA replication *in vitro*.

In this way a number of proteins were isolated and purified using conventional, biochemical techniques. The use of lysates was, however, cumbersome: the bacterial chromosome was present as template, primer, and product and the reactions occurring were complex. To simplify the system, high-speed supernatant fractions (which were devoid of bacterial DNA) were prepared and supplemented with cyclic, single-stranded DNA obtained from a number of phages. Such DNA which, when added to the extract, rapidly complexed with SSB (see Section 2), resembles the DNA on the lagging side of the fork and it was hoped that it would enable the initiation and extension of Okazaki pieces to be studied in isolation from other events. Cyclic DNA has no 3′ OH end to fold back and produce a primer which might confuse the situation (27).

The reactions which occurred were the same as those which occur following infection of *E. coli* with these phages, that is the conversion of a single-stranded molecule to a cyclic duplex known as the replicative form.

The results clearly showed that several phages (including M13, fd, G4, and ϕK) have developed a mechanism of initiation that is unlike that normally involved in the synthesis of Okazaki pieces in the host cell. Thus phages M13 and fd are able to use RNA polymerase to make a short primer at a single, specific site on their cyclic, single-stranded genome and this primer can be extended by holopolymerase III. Such synthesis is sensitive to the drug rifampicin which inhibits *E. coli* RNA polymerase. DNA from phages G4 and ϕK contains a region of secondary structure to which the DnaG primase can bind directly to synthesize a primer which can be extended by holopolymerase III (see Section 5). Furthermore, these phages are able to replicate in several *E. coli* mutants which are defective in bacterial DNA replication.

Only when cyclic, single-stranded DNA from phage ϕX174 was used in the *in vitro* assay was synthesis defective in high-speed supernatant fractions from *dna*B, *dna*C, and *dna*G mutant cells, indicating the involvement of DnaB, DnaC, and DnaG proteins in lagging strand synthesis.

Following the successful isolation of a number of essential proteins, attempts were made to reconstruct synthesis of the ϕX174 DNA lagging strand from purified components. Initial attempts failed, indicating that all the essential components had not been provided. Indeed, this was not surprising as the only proteins present were those for which a mutant was available. The synthesis was successful, however, when the mixture of purified components was supplemented with an extract from wild type cells which was able to provide the unknown components. Further *in vitro* complementation assays brought to light two more essential protein fractions; one of these was sensitive to the sulphydryl group inhibitor N-ethylmaleimide, and the other was insensitive. These two protein fractions were called n (later fractionated into three components; n, n′, and n″) and i by Kornberg's group but they are also known as X, Y, and Z, respectively, by Hurwitz's group. The way in which these proteins combine to produce the primosome has been described in Sections 3 and 5.

Similar approaches have been used to discover the eukaryotic proteins involved in replication (28, 29). The *Xenopus* egg has been used as a rich source of such proteins as it undergoes extremely rapid DNA replication and duplex DNA injected into such eggs is replicated in phase with the cellular DNA. When single-stranded, cyclic DNA is added to extracts it is converted to duplex molecules (in a reaction requiring only DNA polymerase α/primase complex and SSB) and the duplex is assembled into chromatin.

In addition, isolated nuclei have been used to study eukaryotic DNA synthesis *in vitro*. As with a bacterial lysate, this system is complicated by the presence of the cellular DNA, but the use of nuclei and extracts from virally-infected cells provides a system with a much simpler primer/template which is also present at a much higher molar concentration than is true for any host replicon. Nuclei from polyoma infected cells have been used to investigate the nature of the RNA/DNA junction as they are freely permeable to nucleotides (Chapter 1, Section 4). Work with nuclear extracts has allowed the reconstitution of viral replication systems from purified components and is considered further in Chapter 3.

9. Conclusion

This chapter has described the proteins involved at the replication fork and has allowed a model of events to be constructed. This model is applicable to the replication of all duplex DNA molecules, although significant differences are found in the details of the initiation of lagging strand synthesis.

10. Further reading

Liu,L.F. (1984) DNA Topoisomerases, Enzymes that Catalyse the Breaking and Rejoining of DNA. *CRC Crit. Rev. Biochem.*, **15**, 1.

McHenry,C.S. (1988) The Asymmetric Dimeric Polymerase Hypothesis: a Progress Report. *Biochim. Biophys. Acta.*, **951**, 240.

Radman,M. and Wagner,R. (1988) The High Fidelity of DNA Duplication. *Sci. American*, **259**, 24.

11. References

1. Weiner,J.H., Bertsch,L.L. and Kornberg,A. (1975) *J. Biol. Chem.*, **250**, 1972.
2. Lahue,E.E. and Matson,S.W. (1988) *J. Biol. Chem.*, **263**, 3208.

3. Kobori,J.A. and Kornberg,A. (1982) *J. Biol. Chem.*, **257**, 13.
4. Lee,M.S. and Marians,K.J. (1989) *J. Biol. Chem.*, **264**, 14531.
5. Smith,K.R., Yancy,J.E. and Matson,S.W. (1989) *J. Biol. Chem.*, **264**, 6119.
6. Goto,T., Laipis,P. and Wang,J.C. (1984) *J. Biol. Chem.*, **259**, 10422.
7. Kjeldsen,E., Bonven,B.J., Andoh,T., Ishii,K., Okada,K., Bolund,L. and Westergaard,O. (1988) *J. Biol. Chem.*, **263**, 3912.
8. Yang,L., Rowe,T.C. and Liu,L.F. (1985) *Cancer Res.*, **45**, 5872.
9. Avemann,K., Knippers,R., Koller,T. and Sogo,J.M. (1988) *Mol. Cell. Biol.*, **8**, 3026.
10. Spitzner,J.R., Chung,I.K. and Muller,M.T. (1990) *Nucl. Acids Res.*, **18**, 1.
11. Nakai,H. and Richardson,C.C. (1988) *J. Biol. Chem.*, **263**, 9818.
12. Wickner,S. (1977) *Proc. Natl. Acad. Sci. USA*, **74**, 2815.
13. Hirose,M., Yamamoto,S., Yamaguchi,M. and Matsukage,A. (1988) *J. Biol. Chem.*, **263**, 2925.
14. Davey,S.K. and Faust,E.A. (1990) *J. Biol. Chem.*, **265**, 3611.
15. Burgess,M.J. (1989) *Prog. Nucleic Acid Res.*, **37**, 235.
16. Ollis,D.L., Brick,P., Hamlin,R., Xuong,N.G. and Steitz,T.A. (1985) *Nature (London)*, **313**, 762.
17. Joyce,C.M. and Steitz,T.A. (1987) *Trends in Biochem. Sci.*, **12**, 288.
18. Schwartz,H., Shavitt,O. and Livneh,Z. (1988) *J. Biol. Chem.*, **263**, 18277.
19. Maki,H., Maki,S. and Kornberg,A. (1988) *J. Biol. Chem.*, **263**, 6570.
20. O'Donnell,M. and Studwell,P.S. (1990) *J. Biol. Chem.*, **265**, 1179.
21. Kornberg,A. (1988) *J. Biol. Chem.*, **263**, 1.
22. Focher,F., Ferrari,E., Spadari,S. and Hubscher,U. (1988) *FEBS Lett.*, **229**, 6.
23. Weinberg,D.H. and Kelly,T.K. (1989) *Proc. Natl. Acad. Sci. USA*, **86**, 9742.
24. Tsurimoto,T. and Stillman,B. (1989) *EMBO J.*, **8**, 3883; (1990) *Proc. Natl. Acad. Sci. USA*, **87**, 1023.
25. Grosse,F., Krauss,G., Knill-Jones,J.W. and Fersht,A.R. (1983) *EMBO J.*, **2**, 1515.
26. Lasken,R.S. and Goodman,M.F. (1985) *Proc. Natl. Acad. Sci. USA*, **82**, 2189.
27. Kobori,J.A. and Kornberg,A. (1982) *J. Biol. Chem.*, **257**, 13770.
28. Mechali,M. and Harland,R.M. (1982) *Cell*, **30**, 93.
29. Almouzni,G. and Mechali,M. (1988) *Biochim. Biophys. Acta*, **951**, 443.

3

Initiation of replication—single replicons

1. Introduction

There is a regular alternation of DNA replication and cell division that is essential in order to maintain a constant amount of DNA per cell. This controlled replication is found in eukaryotic and prokaryotic cells, and is required also to maintain a constant copy number for plasmids carried in these cells. Viruses, however, generally exhibit run-away replication in which initiation of DNA replication is unlinked to the host cell division cycle; and some plasmids can dramatically change their copy number in response to conditions.

Control over replication is exerted by limiting initiation to a specific location(s) on a chromosome, and to a specific time in the cell cycle. Over the last few years, much has been learnt about the structures of origins of replication, but less is known about what triggers their action. What is clear is that initiation of replication can occur only when events occur which firstly unwind the DNA and also provide a primer for DNA polymerase at the origin.

Primers may be of DNA or of RNA, or even a single, protein-bound nucleotide; and the RNA primers may be synthesized by RNA polymerase or by primase. Primers made by primase are usually very short (Chapter 2, Section 5) but primers made by RNA polymerase may be several hundred nucleotides long.

The DNA duplex may be opened up by transcription, or by the binding of specific proteins which allow helicase action. Opening of the duplex is easier in AT-rich regions, and these are characteristic of origin sequences.

This chapter is concerned with the nature of origins of replication, and how replication of duplex DNA molecules is initiated and controlled. Section 2 provides some of the evidence which shows that origins occur at particular locations on the DNA. Later sections of this chapter investigate the reactions which occur at specific origins in systems where there is (usually) only one origin per chromosome. Chapter 4 considers systems with multiple origins; that is, typically the chromosomes of higher eukaryotes.

2. Methods used to locate origins

2.1 Viral and plasmid origins of replication

All viruses and plasmids so far investigated initiate replication at a single, specific site on their chromosome. (If the primary site is deleted, secondary origins are sometimes activated, and often such secondary origins are used at later stages in infection.)

Such conclusions are easily reached from a study of electron micrographs of the replicating DNA from small linear viruses such as phage T7 (1) (*Figure 3.1***a**). Early in replication, a single 'bubble' is seen, centred on a point 17% from the left hand end of the molecule; and this bubble grows bidirectionally. As the left moving fork reaches the end of the DNA a Y-shaped molecule is formed.

Electron micrographs of cyclic DNA molecules (2) also show a single replicating bubble (*Figure 3.1***b**), but give no information as to its location or whether replication is bidirectional. Often this dilemma can be resolved by cleavage with

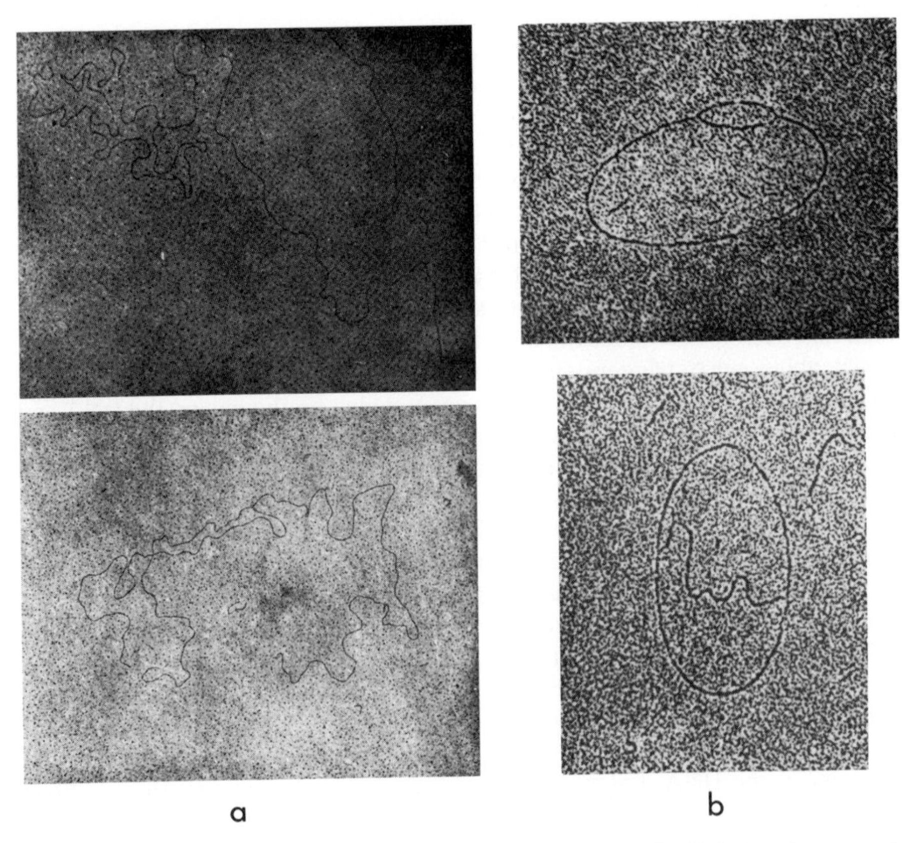

Figure 3.1. Electron micrographs of replicating DNA. **a** T7 DNA showing an early and a later (Y-shaped) replicating molecule. **b** Polyoma DNA showing the typical θ structure. From (2 and 3) with kind permission of the authors and publishers.

a restriction enzyme which cuts the DNA molecule at one site only. Such procedures have shown that, although the site of initiation is unique, replication may be unidirectional (e.g. ColE1) or bidirectional (e.g. SV40 DNA).

Denaturation mapping (3) can be used with larger molecules (e.g. phage lambda DNA), where difficulty may arise in finding a suitable restriction enzyme that will make a single cut.

2.2 The bacterial division cycle

Slowly growing bacteria always undergo fission about one hour after replication starts. As replication takes about 40 minutes there is a period after replication ends before division occurs, and a similar gap may be present between division and initiation of replication (4). In faster growing bacteria, when the frequency of division is increased, the time between initiation of DNA replication and division remains constant at 60 minutes, but mechanisms come into play which allow initiation events to occur twice or even three times per hour. Thus a bacterial chromosome may be undergoing three rounds of replication simultaneously (*Figure 3.2*).

The *E. coli* chromosome is too large to see all of it on a single electron micrograph, but the original experiments of Meselson and Stahl (see Chapter 1, Section 1) were consistent with a single, unique origin and progression of a replication fork along the chromosome once per generation. Other possibilities would not have produced the clear, stepwise appearance of LH and LL duplexes as is shown in *Figure 1.2*.

Further evidence for this comes from a consideration of gene copy number which can be estimated using transformation or transduction assays. Non-growing

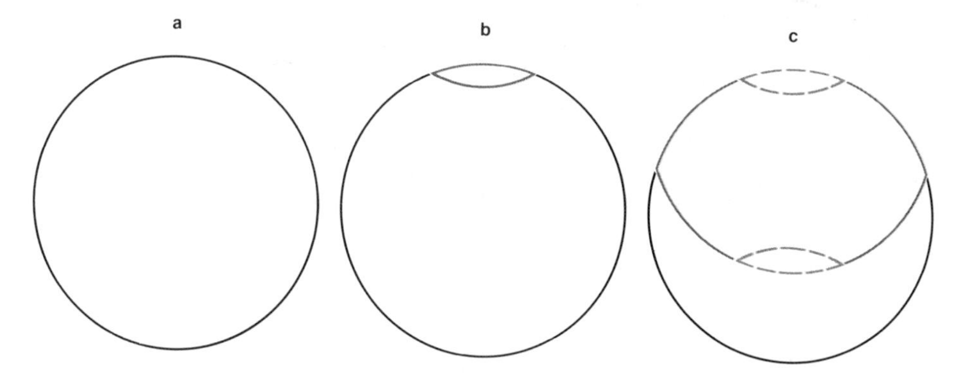

Figure 3.2. Replicating *E. coli* chromosomes. **a** the single, cyclic chromosome is present in a stationary phase cell and is the form present following division of a slowly growing cell. **b** initiation of bidirectional replication at the single origin; this is the form of the chromosome present immediately following division in cells doubling every 50 minutes. **c** in cells doubling every 25 minutes the newly segregated chromosome has reinitiated replication a second time giving 4 copies of the origin for each termination region in every chromosome. Just prior to division each rapidly-dividing cell contains two such chromosomes and, hence, has 8 times as many copies of the region adjacent to the origin as are present in a stationary phase cell.

cells each have a single chromosome and hence a single copy of each gene. Growing cells have a replicating chromosome and hence two copies of genes near the origin but still only one copy of a gene at the terminus of replication. This difference is accentuated in rapidly growing cells where genes near the origin may be present in up to eight copies (*Figure 3.2*). From measurements of the copy number of genes in stationary and rapidly growing cells, it is clear that *E. coli* replicates bidirectionally from a single origin at about 73 minutes, on the genetic map and that termination occurs approximately opposite the origin, at 29 minutes (5).

The size of the *E. coli* chromosome is 3.9×10^6 bp and it takes 40 minutes for one round of replication to be completed. As replication is bidirectional this means the rate of fork movement is about 800 bp per second.

3. Rolling circle replication

Replication of the duplex, RF DNA of small single-stranded phage and of plasmids at conjugation is initiated by a specific endonuclease nicking one strand of the duplex at a specific site (6). The 3′ OH group so generated acts as a primer for DNA polymerase (*Figure 3.3*).

In all cases the endonuclease is phage (or plasmid) encoded and may cut the viral (+) strand of the DNA at a sequence which lies in the coding sequence of the endonuclease gene. In this way, production of the endonuclease is autoregulated, and replication is initiated at a precise nucleotide sequence.

3.1 φX174 RF DNA replication

For replication of phage φX174 RF DNA the endonuclease is the product of gene A and it cleaves between nucleotides 4305 and 4306 in a region of twofold symmetry and becomes bound via a phosphotyrosine link to the 5′ phosphate produced. The 3′ OH terminus binds DNA holopolymerase III and is extended using the 'rolling' (−) strand as template and the (+) strand is cast off, bound to SSB. Unwinding of the DNA is catalysed by Rep, a helicase that shows affinity

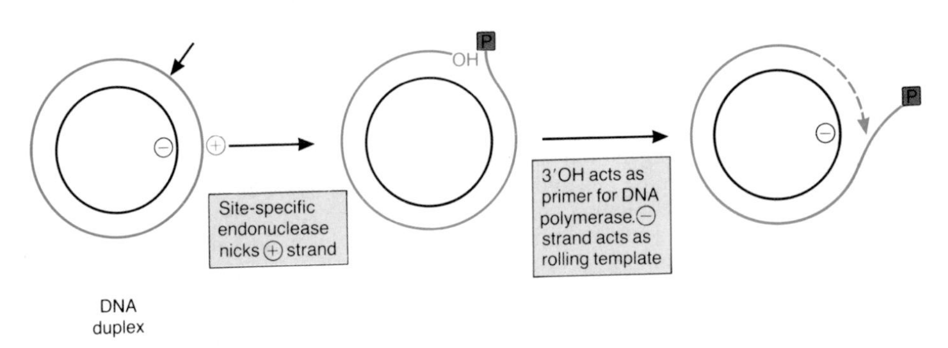

Figure 3.3. Rolling circle replication.

for Protein A (7). In this way Protein A and the 5′ end of the DNA are kept closely associated with the replicating fork (*Figure 3.4*).

Depending on whether replication is taking place early or late in the infectious cycle, one of two reactions may occur to the cast-off (+) strand. Early in infection, when the number of DNA molecules is increasing exponentially, primers are made on the (+) strand by the primosome which has remained associated with the DNA following replication of the DNA from single-stranded to replicative form (Chapter 2, Section 8). Okazaki pieces are made on the (+) strand and a typical replicating fork is produced. When the helicase has unwound one complete circle, the attached Protein A is again brought into contact with its recognition site. In a concerted reaction it first nicks the (+) strand at the point of initiation of replication where the parental and daughter molecules are joined, then becomes transferred to the newly generated 5′ phosphate and finally transfers the original 5′ phosphate to the newly generated 3′ OH terminus (*Figure 3.4*). The products of this reaction are a new rolling circle and a single-stranded circle on which are growing several Okazaki pieces. Once these are completed, the duplex molecule can again initiate RF replication as described above.

Late in infection, the cast-off (+) strand becomes associated, not with SSB, but with the proteins encoded by phage genes B, D, F, G, and H which together form the prohead. These prevent initiation of Okazaki pieces and lead to the packaging of the single-stranded product into the phage coat. This is followed by host cell lysis and release of mature phage particles.

Although the end result of replication of the other phage DNAs is the same they differ in details. Thus in phage fd, the gene II protein (the homologue of Protein A) does not show affinity for Rep and so the looped structures are not formed and there is a requirement for DnaA (see Section 4.1). On the cast-off (+) strand retrograde synthesis can only be initiated when the RNA polymerase binding site (M13, fd) or the primase binding site (G4, ϕK) become exposed, and in the former case this results in the production of an almost complete new (+) strand before (−) strand synthesis is initiated. Late in infection of the filamentous phages (M13, fd) the cast-off (+) strand becomes covered, not by the prohead proteins, but by gene V protein which binds the intact single-stranded cyclic DNA to the cell membrane where gene V protein is exchanged for coat proteins. This is important in helping the new phage particles to leave the host without causing lysis.

3.2 Plasmid transfer

During bacterial conjugation a single strand of plasmid DNA passes from the donor to the acceptor cell. Transfer is initiated by nicking the plasmid at *ori*T and the 5′ end produced passes into the recipient. In F-plasmid containing cells, helicase I is involved in unwinding the two DNA strands (at 1200 bp/sec) leading to transfer of the nicked strand with the *tra*YZ endonuclease attached to the 5′ end (8).

In the donor cell the 3′ OH terminus may act as primer for holopolymerase III to generate a new complementary strand as described for ϕX174 and the

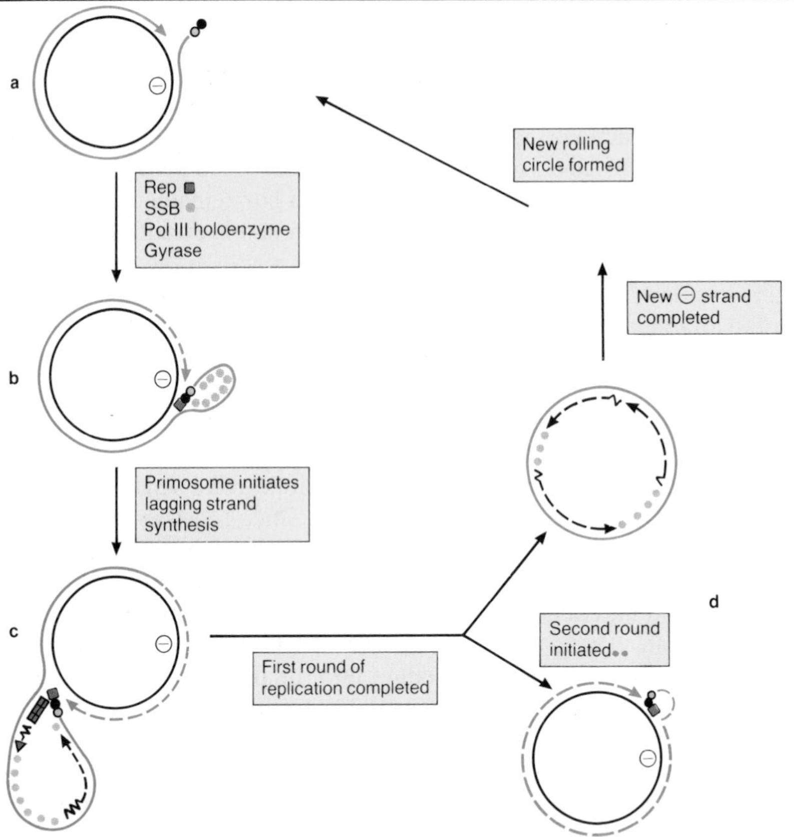

Figure 3.4. ϕX174 RF replication. Protein A nicks the ⊕ strand and binds to the 5′ end **a**. With the aid of Rep and SSB, polymerase III holoenzyme extends the 3′ OH using the uncut ⊖ strand as template **b**. Okazaki pieces are initiated on the lagging side **c**. Replication continues for a complete round when the bound protein A is once again in contact with its recognition sequence and initiates a second round of replication **d**.

rolling circle mechanism. However, in some cases RNA primers are known to be required and ColE1 has n′ binding sites on both strands which would allow primosome formation in both the donor and recipient cell. Some doubt has even been cast on priming by the 3′ OH terminus during ϕ X174 RF DNA replication (8).

Complementary strand synthesis in the recipient can be initiated by a variety of mechanisms: in some cases (F-plasmid) an RNA transcript can act as primer while, with ColE1, formation of a primosome is required. Many other plasmids encode their own, sequence-specific primase making synthesis of the complementary strand independent of DnaG and DnaB.

4. Initiation using primase

4.1 The E. coli *origin of replication (*ori*C)*

The region of the *E. coli* chromosome which is essential for replication has been narrowed down to a stretch of 245 bp (*ori*C) in a 1.3 kb *Hin*dIII fragment. If this region is transferred to a plasmid (whose own origin has been removed) it can serve as an origin, and such constructs are known as *ori*C plasmids. They have been very useful in studies of initiation in *E. coli* as they are much smaller than the whole *E. coli* chromosome.

In *E. coli*, initiation involves using primase to produce a primer at the origin, but the system is complicated by two factors. Primase itself cannot bind to *ori*C and several RNA transcripts are initiated within, or close to the origin (9). This transcription is important for the initiation of replication (transcriptional activation) as it serves to alter the structure of the DNA at *ori*C. This is especially true for the *mio*C transcript which starts to the right of the origin and proceeds leftwards through *ori*C.

Transcription through the origin is essential to allow binding of DnaA to *ori*C. It is also important that the DNA is associated with HU, a histone-like protein which is involved in the packaging of *E. coli* DNA; and that DNA gyrase is available to facilitate unwinding.

By the use of conditional mutants (Chapter 2, Section 8), DnaA has been shown to be essential for initiation of replication of DNA in *E. coli* and many other bacteria and plasmids. The ATP-bound form of DnaA binds to the sequence TT(A/T)T(A/C)CA(A/C)A which ocurs four times within *ori*C (*Figure 3.5*) (10). Overall about 30 molecules of DnaA become bound to *ori*C, acting as a core around which the DNA becomes wrapped. This wrapping causes the DNA to become partially unwound and single-stranded DNA is detected at a very AT-rich region towards the left of the origin. This region, which is made up of three identical 13 bp repeats, also binds DnaA. DnaB helicase, with the help of DnaC, will now bind to both DNA strands in this single-stranded region and enlarge it by unwinding the DNA (*Figure 3.5*). DnaG primase can now bind to the two DnaB complexes and primers are formed on both strands. These primers are extended by holopolymerase III and the DNA formed acts as the leading strand on both sides of the origin; that is, they serve to initiate bidirectional replication (*Figure 3.5*). On both sides of the origin primosome formation occurs leading to initiation of lagging strand synthesis.

DnaA also binds to its own promoter and to the promoters of several genes, including *dam* and *mio*C, whose activity varies with the progression of the cell cycle (11). These genes are thus inactivated as DnaA levels rise in preparation for replication. The level of DnaA is autocontrolled to prevent reinitiation, and it will rise again only following breakdown of the molecules bound to the various promoters.

Prokaryotic DNA is methylated at adenine residues in the sequence GATC by the ubiquitous Dam methylase. On replication, the parental strand retains its methyl groups but the daughter strand remains unmethylated for up to 10 minutes. This hemimethylated state serves to discriminate the two strands in mismatch repair (12).

The distribution of GATC is not random; rather, more than eight sites are clustered in *ori*C (*Figure 3.5*) and these have a function in chromosome segregation. Only hemimethylated *ori*C can bind to certain ill-defined regions

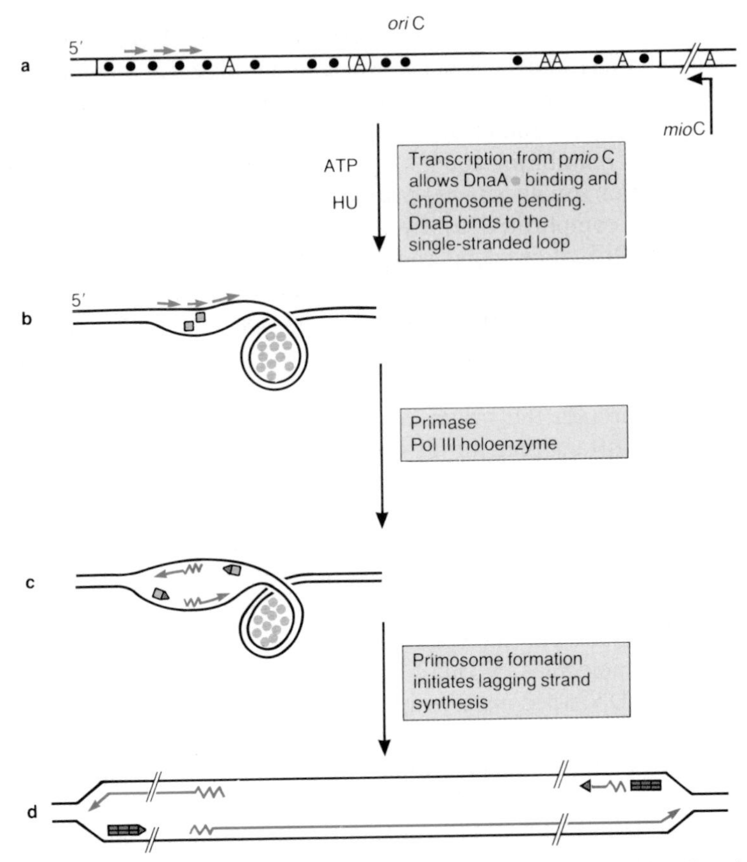

Figure 3.5. The *E. coli* origin of replication—*ori*C. Sites for binding DnaA (A) and GATC (●) sequences are shown along with the three AT-rich repeats (→). Transcription from the *mio*C promoter helps the binding of DnaA which opens up the duplex at the AT-rich region and allows DnaB binding and further opening up of the duplex **b**. Primase and later holopolymerase III initiate bidirectional replication **c** and **d**.

on the cell membrane which grow apart, pulling the two daughter chromosomes with them. When methylation is complete the two chromosomes are released and are partitioned to the two daughter cells (13). Although unmethylated *ori*C plasmids will replicate in Dam^- cells, methylated or hemimethylated plasmids show very inefficient replication in Dam^- cells.

4.2 Initiation of replication of phage lambda DNA

On entry into an *E. coli* cell, linear phage lambda DNA circularizes, by ligating together its 'sticky' ends. Initiation of replication then resembles initiation at *ori*C, except that the host proteins DnaA and DnaC are not required. Instead, virally-coded proteins O and P are used in order to bring about the binding of DnaB helicase. There is no evidence that transcriptional activation is required to facilitate binding of O protein. The O protein binds to four, 18 bp repeats one of which is in its own promoter which leads to autoregulation of synthesis of O. Binding of O leads to the limited opening up of the duplex at an AT-rich region in the origin (*Figure 3.6*) and allows binding of a DnaB-P protein complex which links the single-stranded region to the O protein. This is a stable complex and it remains anchored at the origin in the absence of the three other proteins that are needed, DnaJ, GrpE, and DnaK. These three proteins act to remove O and P from the complex and allow DnaB helicase to move away from, and enlarge the single-stranded region to the right of the origin. DnaJ and DnaK show similarities in sequence to eukaryotic heat shock proteins and may act to facilitate the unfolding of the initial complex (14–16). DnaJ, GrpE, and DnaK play a similar function in plasmid replication (Section 4.3) and, probably, also at *ori*C. Once located on each strand of the single-stranded region (*Figure 3.6*) DnaB associates with DnaG primase. Primers are generated for holopolymerase III, thereby initiating leading strand synthesis for bidirectional replication.

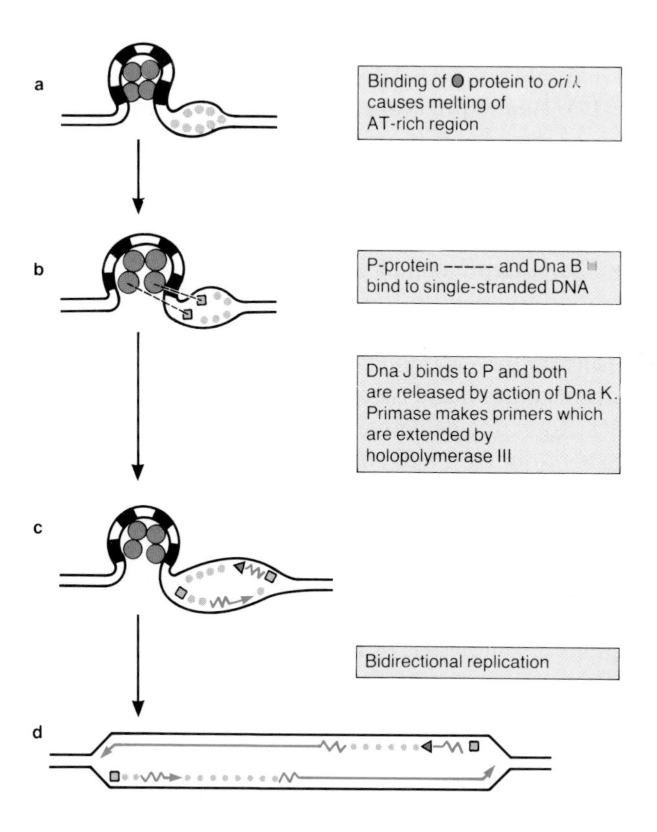

Figure 3.6. Lambda origin of replication. The binding of O-protein causes an unwinding of the duplex at an AT-rich region **a** and allows binding of a P-protein/DnaB complex **b**. Disassembly of the complex by DnaJ and DnaK allows the DnaB helicase to further open up the duplex and DnaG to initiate primers **c** leading to bidirectional replication **d**.

4.3 Low copy-number plasmid replication

Plasmids fall into a number of incompatibility groups, that is, two plasmids from the same group cannot coexist in the same cell. Some plasmids are conjugative (i.e. can bring about their own transfer from a donor to a recipient cell) while other plasmids rely for mobilization on the transfer functions of conjugative plasmids from other groups present in the same cell (Section 3.2.2). The conjugative plasmids (e.g. F, R1, R100) tend to be large (approximately 100 kb) and present in only one or two copies per cell, but RK6 is 38 kb and is present at 13–40 copies per cell. P1 is the prophage form of a temperate phage but resembles low copy-number plasmids in other respects. The origin region of P1 has been inserted into a circular DNA molecule to produce miniplasmid P1 (*Figure 3.7*).

Replication of P1 is dependent, not only on DnaA, but also on the plasmid-encoded RepA protein and on the methylation of five GATC sites in the origin region. RepA, complexed with DnaJ, binds to five 19 bp repeat sequences (GATGTGTGCTGGAGGGAAA) in the *inc*C region of the origin. The last of these sites lies in the promoter for the *rep*A gene, whose transcription is thereby autoregulated (17). Binding of RepA alters the conformation of the DNA and allows binding of DnaA, that is, RepA binding takes the place of the transcriptional activation seen at *ori*C. Once DnaA is bound, replication proceeds in a manner similar to that of *ori*C plasmids (Section 4.1) and phage lambda DNA (Section 4.2) in that a DnaB/C complex binds to a single-stranded region at the origin and the DnaK/GrpE complex dissociates DnaA and allows the DnaB helicase to continue to open up the origin region (18).

RepA also binds to nine sites at the 3′ end of the *rep*A gene (the *inc*A region in *Figure 3.7*). Binding here is tighter than to *inc*C, and acts to remove any RepA from *inc*C and from solution and thereby is able to regulate initiation. However, it appears that a RepA molecule can bind simultaneously to its promoter site (in *inc*C) and to a site in *inc*A and so transcription is not initiated until most of the RepA is degraded. As the host cell grows RepA will be degraded and diluted until the *rep*A gene is derepressed. A burst of synthesis of RepA will then lead to binding to *inc*C, inhibition of further transcription and initiation of replication. RepA will immediately become limiting as the origin replicates and so reinitiation will not occur (19). This is the mechanism which serves to regulate copy number and is an example of the action of the ***autoregulatory model*** (20).

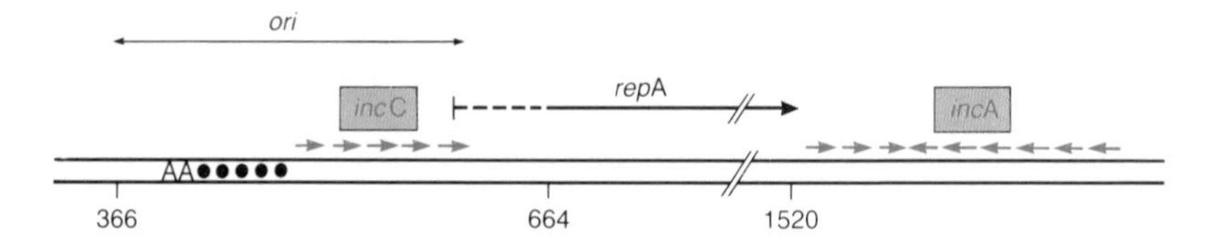

Fig 3.7

Figure 3.7. The origin region of miniplasmid P1. The two incompatibility regions, *inc*A and *inc*C, are shown (→) together with GATCs (●) and DnaA binding sites (A).

A second mechanism exists to control RepA synthesis in plasmids of the FII incompatibility group (e.g. R1 and RK6). The leader region of the RepA mRNA (*cop*T) hybridizes to a constitutive antisense RNA (*cop*A) in a reaction requiring RepA (see also Section 5.2). This acts to inhibit translation as the RepA initiator codon and Shine–Dalgarno sequence are involved in duplex formation. Thus autoregulation of RepA synthesis is exerted at the level of both transcription and translation.

The *Staphylococcal* plasmid pT181 replicates using a rolling circle mechanism (Section 3) in which the essential replication protein (RepC) acts as an origin-specific nuclease (21). Here antisense RNA acts in a different way to limit production of RepC (22). The 5′ flanking region of RepC mRNA normally assumes a hairpin structure which masks a transcription terminator (attenuator). However, an antisense RNA prevents formation of this structure by hydrogen bonding to its upstream arm and this allows the formation of an efficient attenuator structure containing a short duplex region followed by an AU_6 sequence.

By incubating a temperature sensitive mutant of pT181 at the non-permissive temperature, cells can be obtained that retain only one plasmid molecule. On returning to the permissive temperature, the copy number increases rapidly, first overshooting and then returning to the normal value. This is strong evidence for the ***inhibitor dilution model*** (23) for the control of plasmid copy number. This evidence points to antisense RNA as the inhibitor of replication which must be diluted to a certain concentration in order to allow initiation of replication to occur.

4.4 Initiation of papovavirus DNA replication

Initiation requires a virally-coded protein, called T-antigen in SV40 and polyoma. SV40 T-antigen (Tag) binds to three 26 bp regions near the origin of replication. Binding to site I interferes with early transcription, the product of which is Tag (24); that is, synthesis is autocontrolled (*Figure 3.8*). Binding to site II (the only site to lie in the 65 bp origin region) requires phosphorylation of Tag at threonine-124 by the cdc2 protein kinase (25), but is prevented by serine phosphorylation. The activity of cdc2 kinase is autoregulated by phosphorylation and is high in S-phase cells (26), whereas Tag becomes associated with the catalytic subunit of a phosphoserine phosphatase at the end of the G1-phase (27). These observations might suggest that SV40 replication would be limited to the S-phase of the cell cycle. However, another function of Tag is to bind (and inactivate) the negative growth regulators p53 and Rb whose normal role is to restrict cell multiplication (28).

Sites I and III (containing auxiliary regions 1 and 2, respectively) are not in the essential origin region but their presence stimulates initiation of replication about 100-fold. Although the promoter of the Tag gene is partly present in *aux*-1, transcription can be blocked by α amanitin (an inhibitor of RNA polymerase II) without affecting replication if Tag is provided from another template (that is, in *cos* cells).

Site II contains two GAGGC sequences on each strand and binds 12 molecules of Tag. In the presence of topoisomerase I, this causes a local untwisting of the

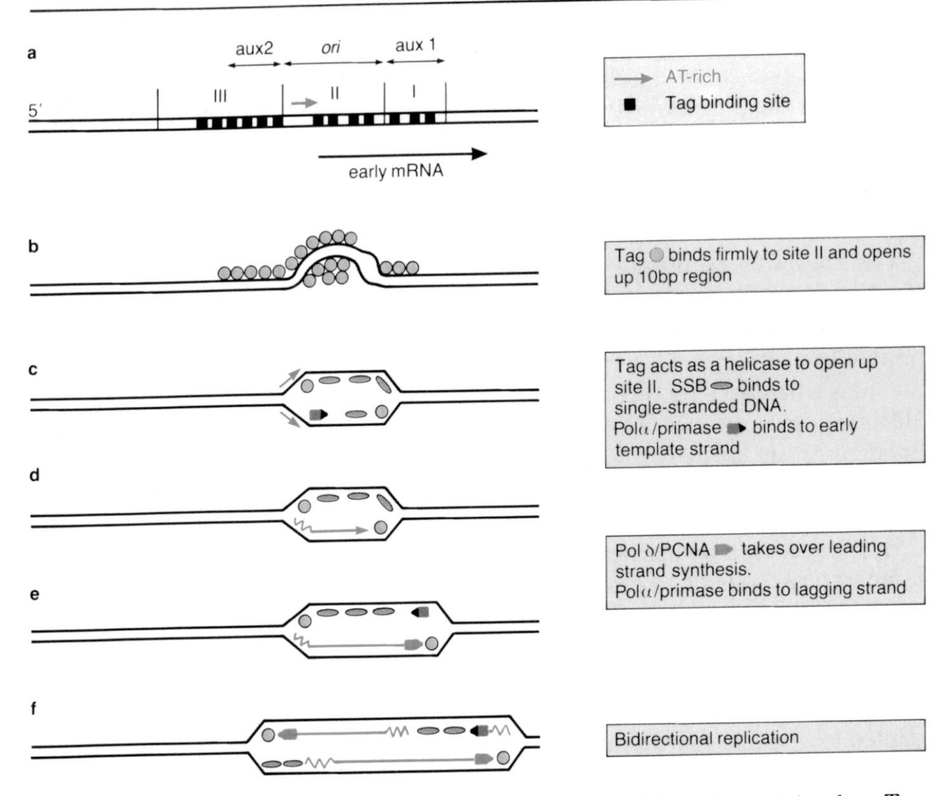

Figure 3.8. The SV40 origin of replication. The essential origin contains four Tag binding sites (■) and AT-rich region (→). The auxiliary regions also contain Tag binding sites (■) **a**. Tag (○) binds firmly to site II (and less firmly to sites I and III) and opens up a 10 bp region of duplex between sites I and II **b**. Acting as a helicase Tag, along with an SSB complex (⬬), opens up the whole origin region and allows polymerase α/primase (□▶) to bind **c** and initiate replication of the early template strand **d**. Polymerase δ/PCNA complex (■▶) takes over leading strand synthesis while polymerase α/primase initiates lagging strand synthesis in site I **e**. This soon becomes the leading strand for bidirectional replication **f**.

DNA starting at the border of sites I and II. Tag now acts as a helicase to produce a 50 bp single-stranded bubble covering most of the origin region; that is, Tag combines the sequence-specific binding property of DnaA with the helicase activity of DnaB. Helicase action is stimulated by a eukaryotic SSB (RF-A). Tag does not bind tightly to single-stranded DNA and the initiating structure could fall to pieces were it not stabilized by other molecules of Tag which are bound to *aux*-1 and *aux*-2 (29). Changes in the phosphorylation status of Tag may be integral to the binding and release of Tag.

The DNA polymerase α-primase complex is able to bind to Tag (in a species-specific manner) at the single-stranded origin region and synthesize a primer at a number of places on the template strand between site I and site II. The primers are initially extended by DNA polymerase α but later the DNA polymerase δ-PCNA complex must take over. There are no initiations on the other strand in the origin region; rather lagging strand synthesis is initiated outside the origin region, but then passes through the origin to become the leading strand for rightward moving synthesis (*Figure 3.8*) (30).

SV40 undergoes run-away replication (see Section 1), but the closely related bovine papilloma virus (BPV) normally replicates only once per cell cycle; that is, BPV replicates as a plasmid with a copy number of about 150 molecules per cell. What controls the initiation of BPV replication to one round per cell cycle while SV40 initiation occurs frequently?

The BPV equivalent of Tag is the E1-protein which has an extra N-terminal domain. Mutants with deletions in this region show run away replication indicating a role for this region in control. Coinfection of cells with SV40 and BPV does not inhibit SV40 run away replication showing the action of the E1 protein cannot be exerted in *trans*. This may be due to failure of E1 to bind to SV40 DNA. Infection of cells with a composite virus containing the SV40 and the BPV genomes leads to controlled replication; that is, the presence of E1 bound to the BPV origin prevents replication from the SV40 origin (31). Whether this involves the physical blocking of replication forks or whether there is some conformational change to the composite DNA molecule which prevents Tag from acting is not yet known.

4.5 Yeast plasmid replication

Some strains of yeast cells contain about 100 copies of a small, 2 micron-long plasmid which contains an origin, enabling it to replicate autonomously (which is therefore called an autonomous replicating sequence or ARS) and a centromere-like element (REP3) required to partition the plasmid evenly to the two daughter cells on division (*Figure 3.9*). The plasmid contains a 599 bp inverted repeat separating two unique regions of approximately equal size. The origin region is a very AT-rich, 11 bp sequence flanked by sequences which are also AT-rich. It is positioned at the boundary of one of the repeat sequences.

This yeast plasmid is of interest because it has been used to search for eukaryotic origins of replication (Chapter 4, Section 4), and because of the manner in which copy number is maintained. Although replication depends only on host proteins, control of copy number, and partitioning depend on the plasmid-coded enzymes, Rep1, Rep2, and FLP.

Together, Rep1 and Rep2 inhibit production of Rep1 which is thereby autocontrolled. Rep1 and Rep2 also act together to repress a third plasmid gene that codes for a recombinase known as FLP. If the copy number of the plasmid falls due to excessive host cell multiplication, the concentration of the Rep1/Rep2 complex falls leading to induction of FLP (and Rep1) (32). FLP catalyses recombination between the two inverted repeats which occur at opposite sides of the plasmid and this leads to inversion of the two halves of the plasmid (*Figure 3.9*). This converts a replicating θ structure into a double rolling circle (Section 3), leading to amplification of the plasmid copy number (i.e. new initiation events are not required to replicate the plasmid). Further recombination events

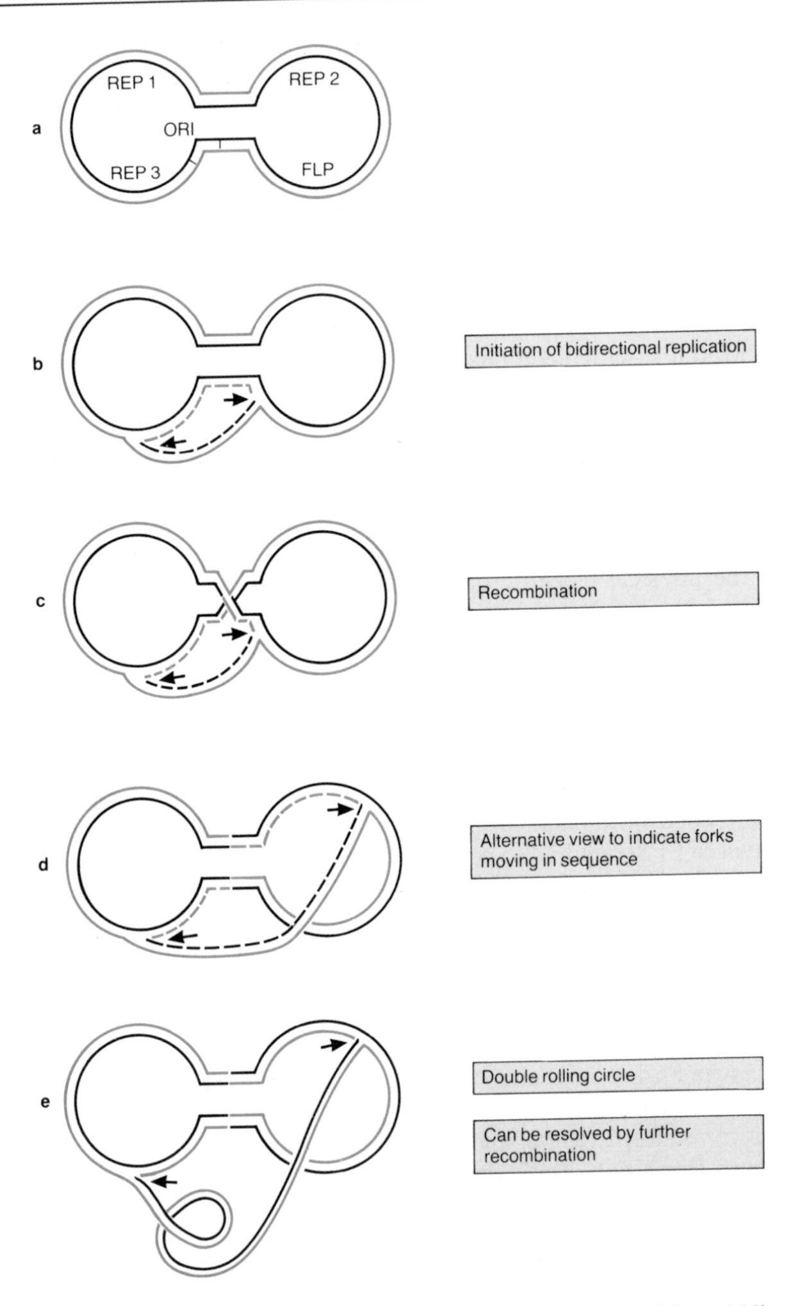

Figure 3.9. Double rolling circle. The yeast 2 micron plasmid initiates bidirectional replication at a region (*ori*) overlapping one end of the duplicated sequence **a**,**b**. Recombination across the redundant sequence means that the two replication forks are now travelling around the cyclic molecule in the same direction **c**,**d**. This structure may continue to replicate to produce single **e**, or multiple copies before further recombination events lead to their resolution.

regenerate θ structures and convert double length concatameric circles into monomers. Amplification of copy number leads to increased levels of Rep1 which repress further FLP production.

5. Initiation using RNA polymerase

5.1 Phage T7 DNA

Initiation of replication of phage T7 DNA occurs 17% from the left hand end of the linear duplex DNA molecule (*Figure 3.1***a**). At this position there are two very strong promoters for the phage-coded RNA polymerase, closely followed by an AT-rich region (*Figure 3.10*). Transcription, initiated at these promoters, terminates in the AT-rich region and the transcripts are extended rightwards by the T7 DNA polymerase (33). This is a complex between the host-coded protein, thioredoxin, the virally-coded polymerase (gene 5) and helicase/primase (gene 4) (see Chapter 2, Section 5). SSB covers the lagging side of the fork until a region of single-stranded DNA containing the sequence (T/G)GGT has been exposed. The helicase/primase then synthesizes a primer of sequence pppACC(C/A) which can be extended leftwards by phage T7 polymerase, back through the origin, to form the leading strand for leftwards fork movement.

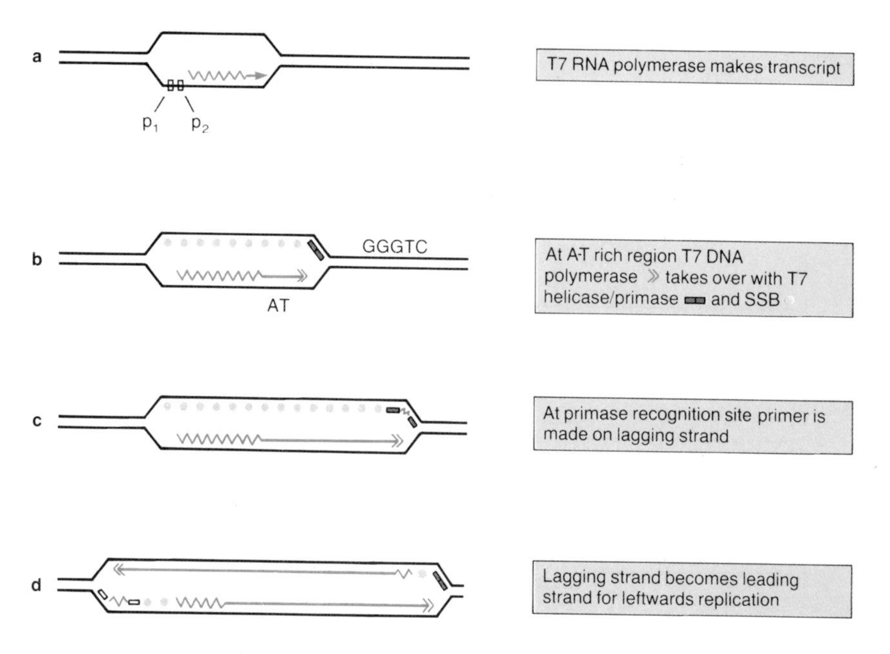

Figure 3.10. The phage T7 origin. 17% from one end of the linear duplex there are two strong promoters for T7 RNA polymerase **a**. Transcription ceases at an AT-rich region and the primer is extended by T7 DNA polymerase (≫) with the help of the T7 helicase/primase (▭) **b**. When the sequence 5′ GGGTC is encountered on the lagging strand a primer is made to initiate leftward replication **c**,**d**.

Meanwhile the helicase/primase has reassociated with the rightwards moving polymerase on the leading side of the rightwards moving fork.

5.2 Initiation of plasmid ColE1 replication

ColE1 is a small, multicopy plasmid from which many of the plasmids used in cloning have been derived. It replicates unidirectionally by means of a θ structure. RNA polymerase initiates transcription at a promoter 555 bp upstream from the origin, to produce a transcript (known as RNA II) that remains base-paired to the template strand in the origin region (*Figure 3.11*). This structure is known as a displacement loop (D-loop) as the RNA-like strand of the DNA is looped out (34). The RNA strand in an RNA:DNA hybrid is sensitive to RNaseH, which introduces several nicks with 3′ OH ends into the RNA. The 3′ OH ends can be extended by *E. coli* DNA polymerase I which has the capacity with its 5′ to 3′ exonuclease (lacking in polymerase III) of removing any surplus, downstream primers. In this way the D-loop is extended to 300 bases, at which point DNA polymerase III holoenzyme takes over.

Lagging strand synthesis is initiated when an n′ binding site becomes exposed downstream of the origin, allowing primosome formation. Lagging strand

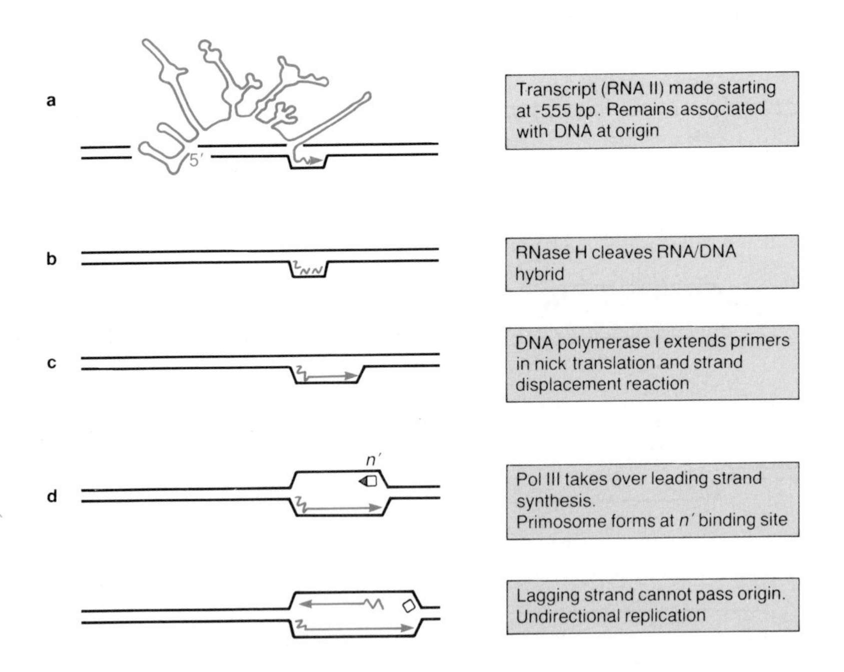

Figure 3.11. The ColE1 origin. 555 bp upstream from the origin a transcript (RNA II) is initiated which forms a complex secondary structure and remains bound to the DNA in the origin region to form a D-loop **a**. RNaseH cleaves the RNA strand in the hybrid to form a series of primers which can be extended by polymerase I **b**,**c**. Polymerase III holoenzyme soon takes over and this exposes an n′ binding site to allow primosome formation on the lagging strand **d**. Synthesis cannot pass back through the origin region.

synthesis does not proceed back through the origin, however, and this may be for one of two related reasons. The RNA polymerase responsible for synthesizing primers is travelling along the DNA in a direction opposite to a leftward moving fork and the former may prevent the fork crossing the origin (see Chapter 4, Section 3). Secondly, lagging strand synthesis is curtailed when it encounters a region of RNA:DNA hybrid duplex that it is unable to pass despite the fact that the template strands for DNA and RNA synthesis are different.

Initiation of replication of ColE1 is controlled by synthesis of a 100 nucleotide long antisense RNA (called RNA I) which is initiated 455 bp upstream from the origin. Both RNA I and RNA II have complex secondary structures containing multiple stem loop regions (*Figure 3.11*), and interaction of the two can occur only in the loops (35). Furthermore, interaction only occurs with nascent RNA II since the mature molecule assumes an alternative structure, presumably involving the loop regions of the incomplete molecule. Interaction, which interferes with the primer function of RNA II, therefore depends on the concentration of RNA I, but there is little information on what controls its synthesis. RNA I can interact with the priming RNA of a number of other plasmids and is the basis for their incompatibility.

A protein encoded downstream of the ColE1 origin regulates the rate of folding of the 3′ end of RNA II. This protein (known as Rom or Rop) increases the time during which RNA II can interact with RNA I and, hence accentuates the inhibitory action of the latter.

The fact that plasmids such as ColE1 are dependent on DNA polymerase I whereas others (e.g. pSC101) use a DnaA protein dependent initiation mechanism has allowed some interesting experiments to be done. A composite plasmid (pSC134) uses only the ColE1 origin and maintains the copy number of ColE1 (36). This shows that, as the host cell grows and inhibitory factors (antisense RNA) are diluted, the ColE1 origin is activated before the pSC101 origin. In pol A^- cells (that make no DNA polymerase I and, therefore, cannot extend the RNA primers produced at the ColE1 origin) the ColE1 origin is inactive and the composite plasmid uses the origin and maintains the copy number of pSC101. Thus, in a composite plasmid where control by inhibitor dilution and autoregulation are both possible, it is the former which normally acts to regulate initiation of replication.

5.3 Mitochondrial DNA replication

The mitochondrial DNA of vertebrates is a covalently-closed, cyclic duplex molecule of about 16 kb. The two strands have a markedly different base composition and are called *H*eavy and *L*ight on the basis of their relative densities in CsCl.

As with papovavirus DNA (Section 4.4), the origin of replication and the start of transcription of both strands occur in the same region of the DNA. However, in the case of mitochondrial DNA replication is unidirectional and highly asymmetric.

The mitochondrial RNA polymerase is responsible for both transcription and

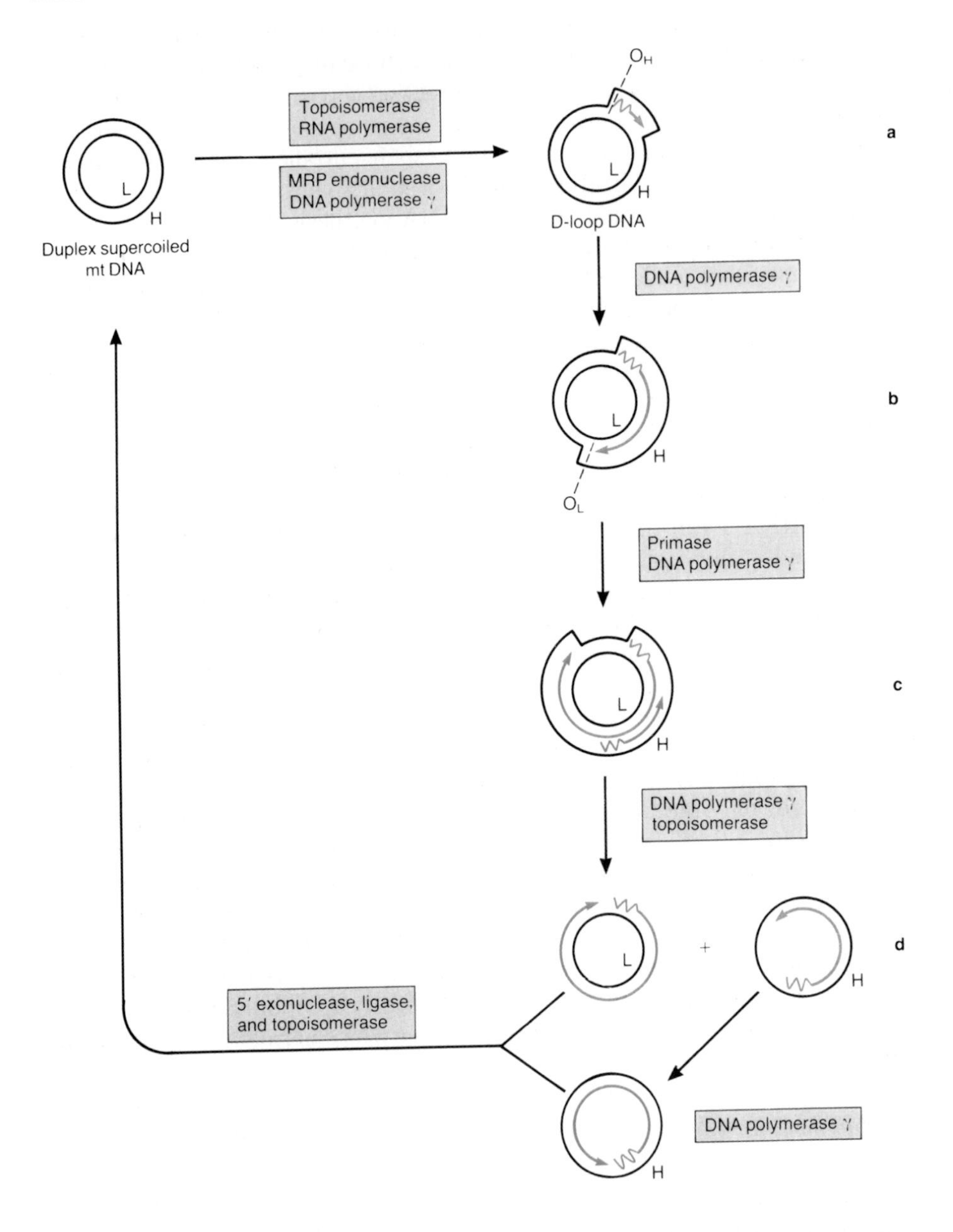

Figure 3.12. A possible scheme for mitochondrial DNA replication. Mitochondrial RNA polymerase synthesizes a transcript which is processed by MRP endonuclease. This provides a primer at O_H for DNA polymerase γ to synthesize a short length of DNA **a** which is later extended by displacement synthesis **b**. When O_L becomes exposed, primase can bind and synthesize a primer which is extended, again by polymerase γ **c**. Asymmetric synthesis leads to separation of the daughter molecules while one is still incomplete **d**.

the synthesis of RNA primers. Mitochondrial RNA polymerases from yeast, *Xenopus*, and human cells consist of a core enzyme (which shows some sequence homology with phage T3 and T7 RNA polymerases) and transcription factors that confer some specificity. Transcription is symmetrical with both strands being transcribed from promoters in the non-coding region of the DNA to give polycistronic transcripts of the whole of the genome. These transcripts are usually processed to give the functional transcripts. Alternatively, the H-strand transcripts may be processed by a specific ribonucleoprotein endonuclease, RNase MRP, which appears to be responsible for cleaving transcripts at a sequence adjacent to a conserved sequence in mammalian mitochondrial DNA (termed CSB II) to give potential RNA primers (37). The 3′ end of the primer is then extended by the mitochondrial DNA polymerase γ (38). This initiation reaction requires unwinding of the duplex and results in the synthesis of a stretch of DNA 450 – 1000 nucleotides long, complementary to the L (light) strand. This structure is known as a D-loop and the newly synthesized piece of DNA may turn over several times before replication proceeds any further (39). This results in a considerable fraction (up to 80% dependent on the species) of the mitochondrial DNA being present with the D-loop structure.

Although the overall strategy for the replication of vertebrate mitochondrial DNA is very widely accepted, the details of individual steps and the enzyme activities actually required for replication remain contentious. *Figure 3.12* shows the likely steps and possible enzyme activities involved in replication. The nascent H-strand in the D-loop may be eventually extended with further synthesis then occurring by displacement of the parental H-strand. Synthesis of the L-strand occurs only when the origin for the light strand (O_L) is exposed. This occurs when about 60% of H-strand synthesis is complete for mammalian mitochondrial DNA, but only after 99% of *Drosophila* mitochondrial H-strand synthesis has occurred. It appears that in the mitochondrial DNA of sea urchins and in kinetoplast minicircle DNA, 'lagging' strand origins may occur at multiple sites.

The O_L sequence is characterized by a stable hairpin structure which is recognized by mitochondrial primase. A 15 – 25 nucleotide primer is made and extended by DNA polymerase γ. Synthesis of both H- and L-strands now occurs but, as synthesis is asymmetric, one daughter molecule is completed while the other is still partly single-stranded (*Figure 3.12*).

Mitochondrial DNA of *Tetrahymena* and *Paramecium* is not cyclic but is a linear duplex. In the former, bidirectional replication occurs from one of two internal origins and is controlled by DNA methylation. In *Paramecium* replication occurs from a crosslinked terminus. The problems associated with the replication of these types of molecule are discussed in Section 4 of Chapter 5.

6. Terminal initiation

The DNA of the phage ϕ29 and of adenoviruses is a linear duplex and replication is initiated at either (or occasionally, both) end(s).

This was shown by labelling adenovirus infected cells with tritiated thymidine for a very short time and isolating those adenovirus DNA molecules which completed DNA replication during the pulse (40). These molecules are labelled only at the replication termini which were localized to the terminal restriction fragments (*Figure 3.13*). This could be interpreted in two different ways. Either replication was initiated in the middle and proceeded bidirectionally (i.e. similar to phage T7 replication—Section 5.1), or replication was initiated at either end and proceeded unidirectionally to the other end. The situation can be resolved as the two strands of the adenovirus DNA duplex can be separated by electrophoresis and it was shown that the labelled fragments hybridized to both strands. This confirmed that replication starts at both ends and not as a central bubble (*Figure 3.13*).

Adenovirus DNA contains an inverted, terminal repeat that varies between 103 and 162 bp in different viral strains. The terminal 18 bp are very AT-rich

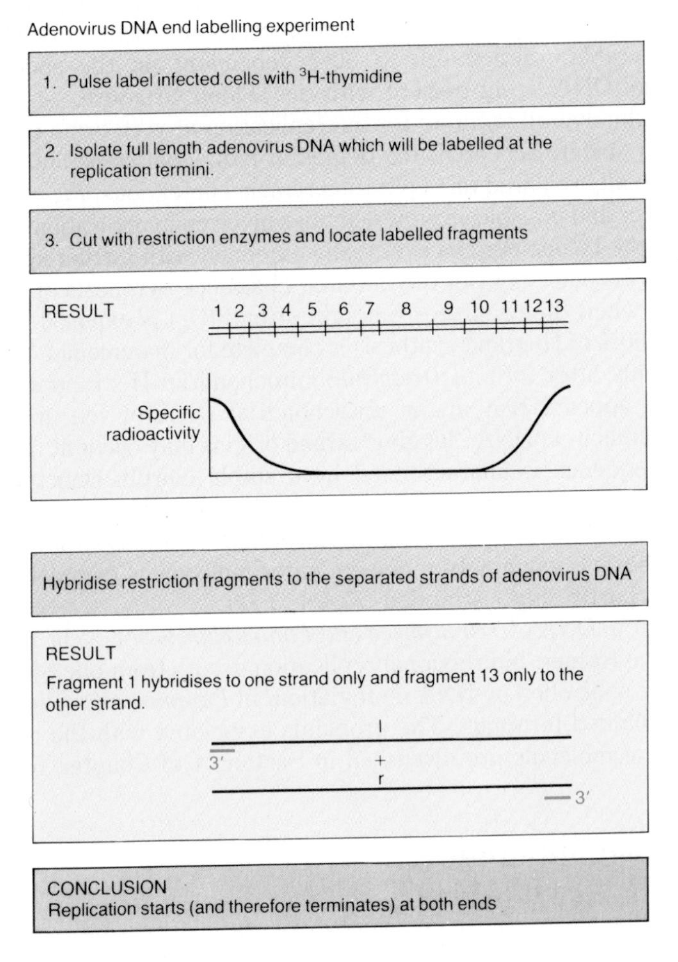

Figure 3.13. Adenovirus DNA end labelling experiment.

and the 5′ end is covalently bound, via a serine phosphate, to a 55 kDa terminal protein (41). Bases 19–39 and 40–51 are involved in binding cellular transcription factors NFI (CTF) and NFIII (OCF-1). In addition to the host cell topoisomerase I (NFII), initiation of replication requires: (i) the virus-coded SSB (72 kDa) which associates with the ends of the duplex to facilitate unwinding; (ii) DNA polymerase (140 kDa); and (iii) terminal protein precursor (80 kDa).

In the presence of NFI, the terminal protein precursor/polymerase complex interacts with dCTP resulting in a dCMP residue becoming covalently linked to a serine residue in the terminal protein precursor (ϕ29 has a dAMP residue

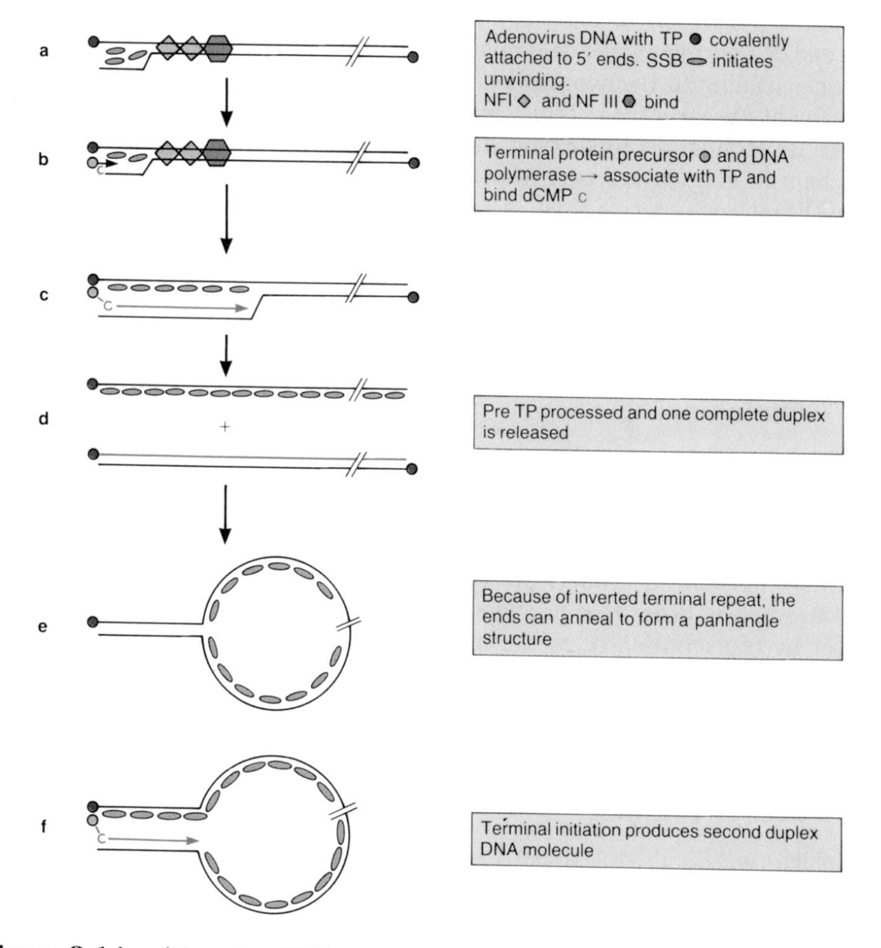

Figure 3.14. Adenovirus DNA replication. Adenovirus DNA has a terminal protein (TP) (●) attached to both 5′ ends **a**. Unwinding of one end is catalysed by SSB, NFI and NFIII in the presence of topoisomerase **a**. Terminal protein precursor (○) and adenovirus DNA polymerase associate with TP and bind dCMP **b** which can serve as a primer for displacement synthesis **c**,**d**. The displaced single-strand can form a panhandle structure **e** which can initiate replication in a manner similar to the original duplex **f**.

attached). This now associates with the 55 kDa terminal protein bound to the 5′ ends of adenovirus DNA where the dCMP serves as a primer to initiate displacement synthesis, catalysed by adenovirus DNA polymerase (*Figure 3.14*). Later the terminal protein precursor is processed to the 55 kDa form to allow separation of the displaced strand from the daughter duplex (41).

As initiation usually occurs at one end only, the product of replication is *one* duplex molecule and a displaced single strand which is covered in SSB and which still retains the terminal protein bound to its 5′ end. As the molecule has an inverted terminal redundancy, the two ends of a single-stranded molecule are complementary and can hydrogen bond with each other to form a 'pan-handle' structure (*Figure 3.14*). The end of the pan-handle has an identical structure to the end of a normal duplex and initiation of replication can occur in an identical manner leading to the synthesis of a *second* duplex molecule (42).

It might appear strange that a phage and a eukaryotic virus have developed a similar strategy of replication. It has been proposed, however, that such a mechanism may have arisen from the rolling circle mechanism of replication of cyclic molecules which have, as an intermediate, a nicked (rather than a linear) molecule with a 5′ bound terminal protein. On the other hand, rolling circle intermediates formed in the later stages of replication of phage lambda DNA (Chapter 5, Section 3) are thought to arise by a failure to terminate at the origin when replication is occurring in a unidirectional manner.

7. Conclusion

In the examples given in this chapter it is clear that replication is initiated at a specific site in each replicon. The major requirement is to provide a primer at the origin and, in most cases, the primer is an RNA molecule. In some instances RNA polymerase may generate a transcript which as a result of site specific cleavage becomes a primer. More often, the primer is synthesized by DNA primase but, in order that primase might bind, the origin requires activation; either by transcription or by the binding of sequence specific proteins which promote the unwinding of the duplex.

Initiation of replication is controlled by a combination of inhibitor dilution and autoregulatory mechanisms, though these are poorly understood at present.

8. Further reading

Bramhill,D. and Kornberg,A. (1988) A Model for Initiation at Origins of DNA Replication. *Cell*, **54**, 915.

Hay,R.T. and Russell,W.C. (1989) Recognition Mechanisms in the Synthesis of Animal Virus DNA. *Biochem. J*, **258**, 83.

Keppel,F., Fayet,O. and Georgopoulos,C. (1988) Strategies of Bacteriophage DNA Replication. In *The Bacteriophages*. Calendar,R. (ed.), **2**, 145, Plenum Press.

Schinkel,A.H. and Tabak,H.F. (1989) Mitochondrial RNA Polymerase: Dual Role in Transcription and Replication. *Trends in Genet.*, **5**, 149.

Scott,J.R. (1984) Regulation of Plasmid Replication. *Microbiol. Rev.*, **48**, 1.

Thomas,C.M. (1988) Recent Studies on the Control of Plasmid Replication. *Biochim. Biophys. Acta*, **949**. 253.

9. References

1. Dressler,D., Wolfson,J. and Magazin,M. (1972) *Proc. Natl. Acad. Sci. USA*, **69**, 998.
2. Hirt,B. (1969) *J. Mol. Biol.*, **40**, 141.
3. Schnös,M. and Inman,R.B. (1970) *J. Mol. Biol.*, **51**, 61.
4. Cooper,S. and Helmstetter.C.E. (1968) *J. Mol. Biol.*, **31**, 519.
5. Master,M. and Broda,P. (1971) *Nature (London)*, **232**, 137.
6. Dressler,D. (1970) *Proc. Natl. Acad. Sci. USA*, **67**, 1934.
7. Brown,D.R., Roth,M.J., Reinberg,D. and Hurwitz,J. (1984) *J. Biol. Chem.*, **259**, 10545.
8. Willets,N. and Wilkins,B. (1984) *Microbiol. Rev.*, **48**, 24.
9. Baker,T.A. and Kornberg,A. (1988) *Cell*, **55**, 113.
10. Zyskind,J.W. and Smith,D.W. (1986) *Cell*, **46**, 489.
11. Georgopoulos,C. (1989) *Trends in Biochem. Sci.*, **5**, 319.
12. Modrich,P. (1987) *Ann. Rev. Biochem.*, **56**, 435.
13. Ogden,G.B., Pratt,M.J. and Schaechter,M. (1988) *Cell*, **54**, 127.
14. Alfano,C. and McMacken,R. (1989) *J. Biol. Chem.*, **264**, 10709.
15. Zylicz,M., Ang,D., Liberek,K. and Georgopoulos,C. (1989) *EMBO J.*, **8**, 1601.
16. Liberek,K., Osipiuk,J., Zylicz,M., And,D., Skorko,J. and Georgopoulos,C. (1990) *J. Biol. Chem.*, **265**, 3022.
17. Pal,S.K., Mason,R.J. and Chattoraj,D.K. (1986) *J. Mol. Biol.*, **192**, 275.
18. Wickner,S.H. (1990) *Proc. Natl. Acad. Sci. USA* **87**, 2690.
19. Chattoraj,D.K., Mason,R.J. and Wickner,S.H. (1988) *Cell*, **52**, 551.
20. Sompayrac,L. and Malloe,O. (1973) *Nature New Biology (London)*, **241**, 133.
21. Khan,S.A., Murray,R.W. and Koepsel,R.R. (1988) *Biochim. Biophys. Acta*, **951**, 375.
22. Novick,R.P., Iordanescu,S., Projan,S.J., Kornblum,J. and Edelman,I. (1989) *Cell*, **59**, 395.
23. Pritchard,R.H., Barth,P.T. and Collins,J. (1969) *Symp. Soc. Gen. Microbiol.*, **19**, 293.
24. Hay,R.T. and DePamphilis,M.L. (1982) *Cell*, **28**, 767.
25. McVey,D., Brizuela,L., Mohr,I., Marshak,D.R., Gluzman,Y. and Beach,D. (1989) *Nature (London)*, **341**, 503.
26. Brautigan,D.L., Sunwoo,J., Labbe,J.-C., Fernandez,A. and Lamb,N.J.C. (1990) *Nature (London)*, **344**, 74.
27. Virshup,D.M., Kauffman,M.G. and Kelly,T.J. (1989) *EMBO J.*, **8**, 3891.
28. Prives,C. (1990) *Cell*, **61**, 735.
29. Gutierrez,C., Guo,Z.-S., Roberts,J. and DePamphilis,M.L. (1990) *Mol. Cell. Biol.*, **10**, 1719.
30. Prelich,G. and Stillman,B. (1988) *Cell*, **53**, 117.
31. Roberts,J.M. and Weintraub,H. (1988) *Cell*, **52**, 397.
32. Som,T., Armstrong,K.A., Volkert,F.C. and Broach,J.R. (1988) *Cell*, **52**, 27.
33. Richardson,C.C. (1983) *Cell*, **33**, 315.
34. Masukata,H. and Tomizawa,J. (1984) *Cell*, **36**, 513.
35. Polisky,B. (1988) *Cell*, **55**, 929.
36. Cabello,F., Timmis,K. and Cohen,S.N. (1976) *Nature (London)*, **259**, 285.
37. Bennett,J.L. and Clayton,D.A. (1990) *Mol. Cell. Biol.*, **10**, 2191.
38. Insdorf,N.F. and Bogenhagen,D.F. (1989) *J. Biol. Chem.*, **264**, 21498.
39. Clayton,D.A. (1982) *Cell*, **28**, 693.
40. Weingartner,B., Winnacker,E.L., Tolun,A. and Pettersson,U. (1976) *Cell*, **9**, 259.
41. Nagata,K., Guggenheimer,R.A., Enomoto,T., Lichy,J.H. and Hurwitz,J. (1982) *Proc. Natl. Acad. Sci. USA*, **79**, 6438.
42. Stillman,B.W. (1983) *Cell*, **35**, 7.

4

Multiple origin systems

1. The eukaryotic cell cycle and replicons

Dividing animal cells pass around the *cell cycle*, which lasts for about 20 hours. After mitosis (M) which takes about one hour there is a gap (G1-phase) of maybe 10 hours before DNA replication (S-phase) starts. S-Phase lasts for 6–8 hours and there is a second gap (G2-phase) before the cell divides again. The cells of lower eukaryotes behave differently and, in particular, the budding yeasts do not undergo mitosis but, rather, one copy of the DNA segregates to a region which buds from the mother cell to form the daughter cell.

Electron micrographs of replicating DNA (1) show a regular, tandem array of bubbles of double-stranded DNA along the duplex molecule (*Figure 4.1*), indicative of multiple origins of replication. The region of DNA replicated from a single origin is called a replicon, and so it is clear that the cellular DNA of eukaryotes (unlike the DNA of viruses, plasmids, and bacteria) contains multiple replicons. What is not clear from the electron micrographs is whether the origins of replicons occur in precise locations along the DNA, and whether replication is uni- or bidirectional. The first of these questions will be considered further in Sections 2–6.

The second question was resolved by studying DNA fibre autoradiographs obtained after labelling synchronized cells with tritiated thymidine. After a short pulse with the radioactive label, the cells are lysed in such a way that the DNA fibres are laid out in parallel arrays along a glass slide. An autoradiograph shows short runs of grains, each run resulting from the labelling of a single replicon, and this confirms the conclusions drawn from electron microscopy (2). When the initial radioactive pulse is followed by a short chase, during which the specific activity of the tritiated thymidine is reduced, the resulting autoradiograph shows runs of grains which are of reduced intensity at both ends. This shows that the replicons were growing in both directions from a single origin. A diagrammatic representation of what is happening is given in *Figure 4.2*.

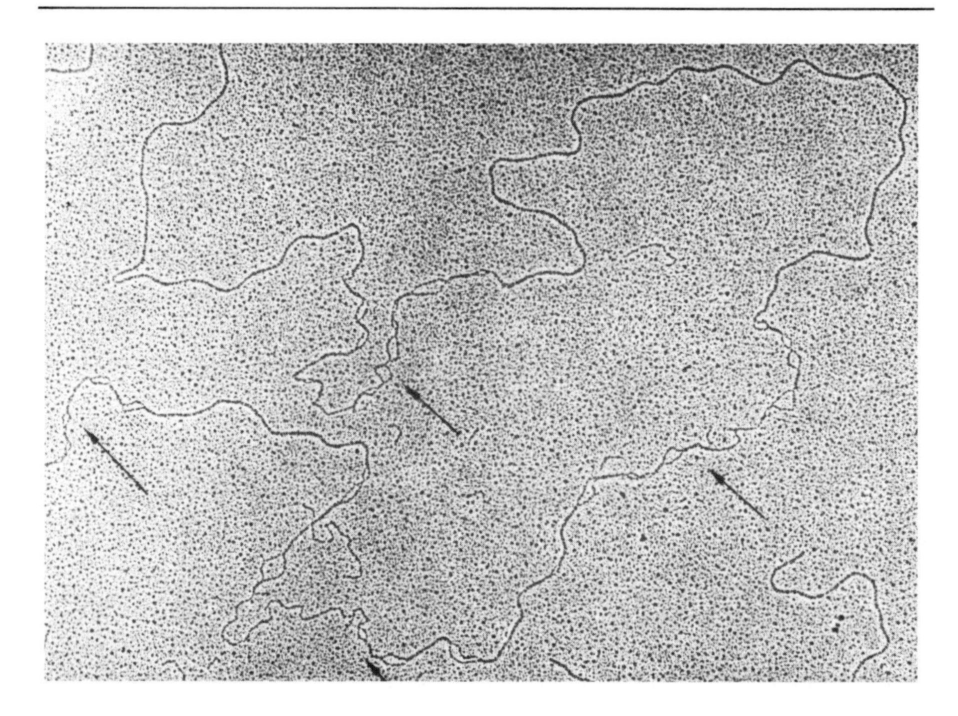

Figure 4.1. Electron micrograph of a replicating *Drosophila* chromosome. Several replication bubbles are seen in tandem array. From (1) with kind permission of the author and publisher.

From measurements of the distance between the centres of adjacent runs of grains, it is possible to come to a value of 40 – 200 kb for the sizes of replicons in tissue culture cells. This is small compared with the size of the single replicon of *E. coli* (3.9×10^6 bp). Knowing that there are about 5×10^9 bp of DNA per mammalian cell it can be calculated that there are about 90 000 replicons per nucleus. Very different values are obtained when similar experiments are performed with cells from certain body tissues. For example the average size of a replicon in a newt spermatocyte is a 1×10^6 bp, whereas in the very rapidly dividing embryonic cells of *Xenopus* and *Drosophila* the average replicon is only 15 000 bp long.

By measuring the length of the run of grains after pulses of different duration, a rate of fork movement of about 50 bp per second is obtained, which is considerably slower than the rate in bacteria (800 bp per sec; see Chapter 3). To account for an S-phase which lasts for 6 – 8 hours, an average of 3000 replicons must be active simultaneously in a cell. A number of procedures have shown that the timing of replication of a replicon in S-phase is not a random process and this is considered further in Section 8. Suffice it to say for now, that adjacent replicons replicate together, such that specific, large regions of a chromosome replicate early in S-phase, while other regions replicate late in S-phase.

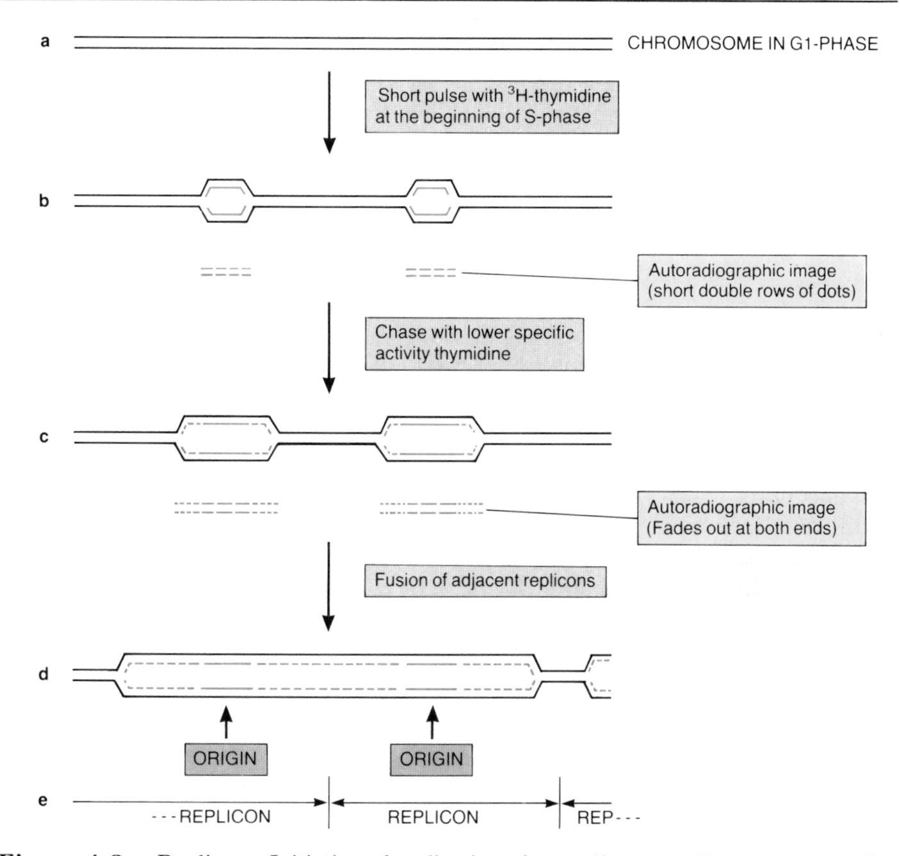

Figure 4.2. Replicons. Initiation of replication of two adjacent replicons occurs at the beginning of S-phase in the presence of tritiated thymidine; the fibre autoradiograph pattern is shown below the diagram in **b**. As replicons grow (now in the presence of lower specific activity thymidine) the rows of grains are seen to fade away at both ends **c**. Eventually adjacent replicons fuse **d**. In **e** is shown the extent of the replicons.

The questions which require an answer are; firstly have all the origins of replication the same or related sequences, or do they fall into a number of classes, or are they all different and secondly what regulates the time of initiation of individual replicons?

2. The origin of replication of repeated genes

Certain genes are present in multiple copies in eukaryotic cells. Thus the rRNA genes are present as several hundred adjacent copies in sea urchin DNA and in multiple copies in the DNA of *Drosophila* and yeast. Another example is the gene for dihydrofolate reductase (*dhfr*) which is amplified up to a hundred thousand times in mammalian cells treated with methotrexate. From a study of electron micrographs of DNA from nuclei containing such genes and from an analysis of which restriction fragments are labelled first when cells initiate a round of replication in tritiated thymidine it is clear that such genes have a specific origin which occurs once per repeat (3). Thus the origin for replication of rDNA occurs in the non-transcribed spacer region (*Figure 4.3*).

3. Replication and the direction of transcription

From a closer study (see Section 5) of the yeast rDNA replicons it is clear that not all origins are used at each S-phase and that replication, although starting in a bidirectional manner, quickly becomes unidirectional. The origin is in a region which is not being transcribed and, hence, replication forks moving in either direction will not, initially, come into contact with a transcription complex. Very soon, however, the leftward fork will come into the 3′ region of the gene and will encounter transcription complexes moving in the opposite direction. This terminates fork movement (4, 5). Rightward moving forks will travel along a transcribed region in the direction of transcription and this does not interfere with fork movement (although the rates of replication and transcription must be synchronized). The rightward moving fork will thus progress, perhaps over several potential replicons, until it encounters a stalled, leftward fork when the two replicons will fuse (*Figure 4.3*).

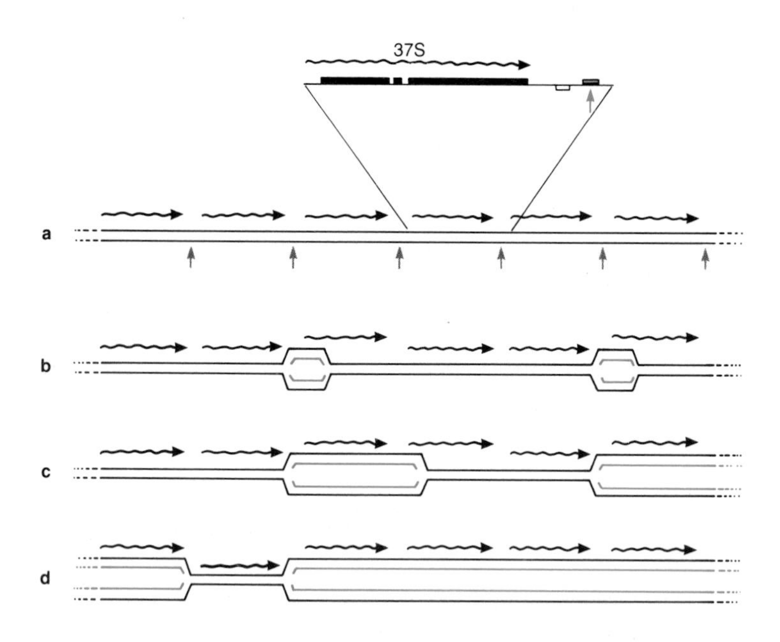

Figure 4.3. Yeast rDNA replicons. Potential origins occur in the non-transcribed spacer of every repeat (↑) **a**, but only some are used. Although bidirectional replication is initiated, leftward synthesis stops when the replication machinery encounters a counter-moving transcription complex **b**. Replicons continue to grow asymmetrically **c** and eventually fuse **d**.

This problem of a replication fork being stalled or derailed by an oncoming transcription complex is abviously most acute for genes, such as that for rRNA, which are frequently transcribed. It may be an important consideration for the location of origins of replication. Thus the origins of replication of the DNA of SV40, mitochondria, and adenovirus are all located at the only or major transcription start site, ensuring that no clash occurs between transcription and replication. Furthermore 66% of *E. coli* genes (and 85% of frequently transcribed genes) are transcribed in their direction of replication. The fact that this figure is not 100% indicates that contra-transcription is not a complete block to replication, and some helicases (e.g. that coded for by the *dda* gene of phage T4) (6) have the ability to displace transcription complexes and allow passage of the replication fork.

Histone H5 is synthesized only in developing avian erythrocytes where the gene is replicated in the direction of transcription. In other avian cells, where the gene is not expressed, a different origin of replication is selected resulting in replication in the anti-transcriptional direction (7). Similarly, a c-*myc* gene translocated next to an IgA heavy-chain constant region is activated on genome rearrangement. Initially it is replicated from an origin 3′ to the inactive gene. Following rearrangement of the antibody gene, a new origin, 5′ to c-*myc*, is activated and the gene is expressed (8).

Thus the problems posed by transcriptional termination of replication may be overcome by selecting alternative origins, and this may involve genome rearrangements.

4. Autonomous replicating sequences (ARS)

Removal of the origin (ARS) from the yeast 2 micron plasmid leads to loss of the plasmid from the cell. If the ARS is replaced by a series of random fragments of DNA from the yeast genome, some of the resulting plasmids regain the property of autonomous replication, that is, the inserts in these plasmids contain host cell origins of replication (9).

A comparison of the sequences of a number of ARS indicate that they are all very AT-rich regions with a conserved, 11 bp consensus sequence of (A/T)TTTAT(A/G)TTT(A/T). Small mutations in this sequence can eliminate ARS function but, similarly, ARS function may be easily generated even in prokaryotic DNA by single nucleotide changes (10).

Cloning mouse sequences into the 2 micron plasmid shows that similar sequences are present in mammalian genomes. It does not show, however, that such regions are origins in mouse cells; only that they (together with certain prokaryotic sequences) can act as origins in yeast.

Attempts to maintain autonomously replicating plasmids in mammalian cells have met with little success as the plasmid DNA quickly becomes integrated into the host chromosome (see, however, ref. 11). An exception to this is a plasmid containing the origin of replication (a 65 bp palindromic region) of Epstein – Barr virus (EBV). Additional requirements for retention of the DNA as a plasmid in this system are firstly, a virally-coded nuclear antigen (EBNA-1) together with its binding site (a tandemly repeated 30 bp sequence); and secondly, a selectable marker to prevent loss of the plasmid. EBNA-1 can be provided in *trans* and does not need to be coded for by the plasmid. By removing the plasmid origin and replacing it with random pieces of the human genome a heterogeneous collection of plasmids was obtained which could replicate autonomously in cultured human cells in the presence of EBNA-1. These plasmids contain presumptive human chromosomal origins in their inserts, but have not yet been characterized in detail (12).

5. Other methods of locating origins

5.1 Gel electrophoresis

Non-linear (i.e. Y-shaped or 'bubbled') restriction fragments do not migrate on gel electrophoresis at the same position as their linear equivalents (13). Furthermore, if a second dimension is run under denaturing conditions, the composition of the anomalously migrating fragments can be analysed (14). From a study of the migration of restriction fragments from replicating yeast DNA molecules, two groups were able to confirm that most ARS elements really are origins of replication, although some appear to be replication termini. The two approaches are slightly different but, by using specific, labelled probes, individual genes can be investigated. Such studies, however, also led to the conclusion that some ARS elements, active in plasmids, are not used as origins in their normal, chromosomal location and that other ARS elements are not used at every S-phase (Section 3). The reasons for this are not clear, but have been discussed by Umek *et al.* (15). Two-dimensional gel electrophoresis can be used to analyse origins of repeated genes in higher eukaryotes, but the signal strength is too weak to allow analysis of single-copy genes. Polymerase chain reaction (PCR) amplification techniques may allow this problem to be overcome.

5.2 Nuclease sensitivity

Handeli *et al.* (16) have identified origins by taking advantage of the finding that parental nucleosomes associate preferentially with the leading strand at the replication fork, and that, in the absence of protein synthesis, the lagging strand is naked and hence more sensitive to nucleases (see Chapter 6, Section 3). Origins and termini can be located by mapping diverging and converging protected regions, respectively.

5.3 Newly-initiated-origin DNA

When cells are labelled with bromodeoxyuridine for 5 – 10 minutes, the nascent DNA fragments can be isolated and separated according to size on alkaline sucrose gradients and then precipitated with anti-BrdU antibodies. Pairs of oligonucleotide primers can be used in the polymerase chain reaction (PCR) to amplify regions around the putative origin and the sequences present only in the smallest fragments will be those nearest the origin of replication (17). Moreover, from a knowledge of the size of these smallest fragments and the location of the PCR primer, the distance from the primer to the origin can be calculated, assuming bidirectional replication. Although not yet used extensively, this approach, which uses PCR to allow detection of minute amounts of material, should allow the origins of single copy genes in vertebrates to be detected.

5.4 Chemical crosslinking

Short nascent strands of DNA can be extruded from Y-shaped replicating molecules by heating and the proportion of short fragments can be enhanced by *in vivo* crosslinking of the parental strands with psoralen (18). On the assumption that the shortest fragments come from nearest the origin, they have been used as probes to determine which restriction fragment of a cloned gene includes the origin region. Results obtained by this method appear consistent with other methods, but again it should be possible to improve the sensitivity and resolution by using PCR amplification methods.

5.5 Density labelling

The replication origins of the histone H5 gene and of c-*myc* (Section 3) have been located using an *in vitro* runoff assay. Isolated nuclei were treated with a restriction enzyme (A) and incubation continued in the presence of BrdUTP. DNA was isolated, cleaved with a second restriction enzyme (B), and fractionated according to density. Few initiations occur *in vitro* and hence the most heavily labelled fragments will be those at termination sites (or at sites of cleavage of enzyme A).

5.6 Hybridization

Regions of DNA near an origin will replicate before those far from the origin, and, as the rate of replication is less than 2 kb per minute (Section 1), it is possible in some cases to locate an origin roughly by hybridizing labelled DNA from synchronized cells to a set of restriction fragments (19).

6. The *Xenopus* egg

Injection of double-stranded DNA into the eggs of *Xenopus* leads to the replication of that DNA throughout cleavage, in concert with the DNA of the host cell, that is, the injected DNA replicates once per S-phase. Any DNA, from eukaryotic or prokaryotic source, will replicate in this system indicating no preference for a putative replication origin.

Replication in cleaving amphibian and insect embryos occurs more rapidly than in any other known system (20). The large amount of DNA present in a frog egg will duplicate itself in about 20 minutes with every cell division. A study of the rate of fork movement and replicon size in amphibian embryos shows that a high rate of DNA replication is achieved by reducing the average size of a replicon to 5 μ (about 15 000 bp), that is, many more origins are active in eggs than are active in somatic cells (see Section 1).

From a consideration of the evidence that not all origins are used in each S-phase; that the time of activation of origins differ; and that extra origins are used in amphibian embryos, it can be concluded that potential origins must interact with a factor or factors in order to be activated and that the abundance of factors differs between cells and throughout S-phase. The more factors that are available, the less discrimination occurs as to what sequence can serve as an origin. Thus in *Xenopus* eggs, factors are abundant and almost any, slightly AT-rich, sequence can and does serve as an origin. This theme is considered further in the next section, but it should be borne in mind that there is another feature which distinguishes the early *Xenopus* embryo; there is only a minimal amount of transcription and so there will be no significant hindrance to the passage of replication forks.

7. Matrix attachment sites

The size of replicons is similar to the size of the loops of DNA which can be seen in the electron microscope when metaphase chromosomes or isolated nuclei are treated in such a way that the histones are released. Moreover, the size of these loops of dehistonized DNA vary in parallel with replicon size, being smaller during cleavage of amphibian eggs (21).

The ends of the loops are held together on the chromosome scaffold or nuclear matrix by protein, and can be isolated by treatment of dehistonized chromatin with nucleases. These nucleases digest the free DNA in the loops, leaving only the DNA that is protected by the scaffold proteins. DNA, pulse labelled with tritiated thymidine either at the beginning of S-phase or at any time throughout S-phase, is found to be enriched in the matrix-associated fraction. This indicates that both initiation and continuing replication of DNA occurs in close association with the matrix (22). Models have been proposed which can explain how replicating loops of DNA are pulled through a matrix bound ***replisome*** or complex of enzymes involved in replication (*Figure 4.4*). Such models assume that a loop is part of two adjacent replicons, and that initiation may occur in one S-phase at what was the site of termination in the previous S-phase. Indeed, termination has been found to be associated with matrix-attachment sites (16) and the idea is not inconsistent with the finding that some ARS elements appear to be termination sites rather than initiation sites (Section 5). On the other hand, it is possible that a reorganization occurs at, or shortly after, mitosis to release termini from the matrix and to reattach origins to matrix-bound replisomes.

Although these results indicate that all regions of DNA can, at some time, be associated with the matrix, they are not inconsistent with other results that show that particular sequences (e.g. topoisomerase II binding sites) are also

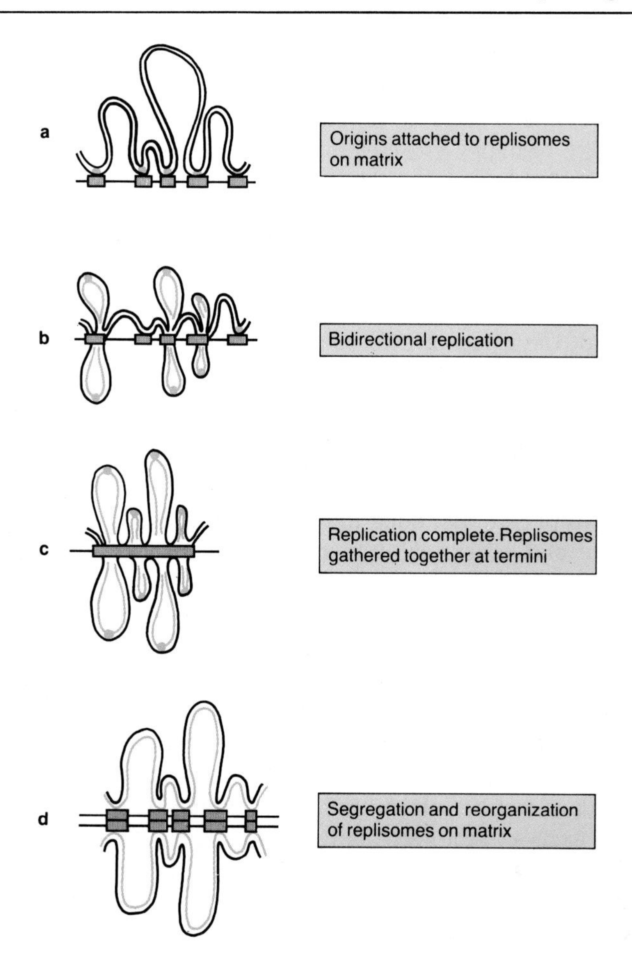

Figure 4.4. Matrix associated replication. Origins attach to replisomes which form part of the nuclear matrix **a** and bidirectional replication leads to formation of paired loops **b**. When replication is complete the replisomes must all be gathered together at the replication termini **c**. Segregation and reorganisation of the origin/replisome complexes must occur before the next S-phase **d**.

associated preferentially with the matrix. A replicon takes only about 15 minutes to replicate and, for the rest of the cell cycle, a specific region of it may be firmly associated with the matrix. Indeed, it is the region which is matrix associated outside of S-phase which is going to be most enriched in origin and termination sequences.

The conclusion that arises from this section is that, in eukaryotes, initiation requires the association of an origin of replication with a matrix associated replisome. The number of replisomes available may be in excess, in which case any AT-rich sequence may be able to act as an origin and this is the situation in amphibian eggs. More usually, however, the number of replisomes is limiting and not all potential origins can bind to the matrix. There will, therefore be a hierarchy of origins. Some which have a high affinity for the replisome will bind readily and their replicon will be replicated early in S-phase. Others will have to wait until a site becomes available and these will replicate later in S-phase. The longer the wait, the greater the chance that a given stretch of DNA will be replicated from a distant origin, and so some potential origins will never be used.

We are still not in a position to define what makes one origin better than another, but effectiveness may be controlled largely by their ease of unwinding (23, 24). Some of the issues are discussed further in Section 8.

8. Temporal replication

Normally genes replicate just once in each S-phase, but some genes replicate early in S-phase and others late. In general, transcriptionally active genes replicate early and the inactive X-chromosome of female mammals is characteristically late replicating (25). Is it likely that transcription of a region of DNA allows or even facilitates initiation of replication; or is the reverse true, that is, replication is required to activate transcription?

A number of studies have shown that the timing of replication of a gene can change following activation (19, 26) and on transfecting the *Xenopus* β-globin gene into human cells only constructs capable of replication were transcribed (27). These findings can be interpreted by assuming that certain transcription factors can gain access to the DNA only at replication. They are not consistent with a requirement for RNA synthesis in order to initiate replication (Chapter 3, Section 5).

A common requirement for transcription and replication is the binding of specific factors and an opening up of the DNA duplex. Indeed, at least one of the common transcription factors (CTF) has been shown to be the same as a factor (NFI) which is required for initiation of replication of adenovirus DNA (28) and which stimulates SV40 DNA replication 20-fold (29). Another factor (NFIII) required for initiation of adenovirus DNA replication (Chapter 3, Section 6) also binds to promoter elements of a number of different genes (30). This might support the view that the observed relationship between transcription and replication arises simply through a requirement for common factors and that when these are available for transcription of a particular gene they are simultaneously available for its replication.

Further evidence for this conclusion arises from the fact that transcriptionally inert genes are packed into a highly condensed chromatin structure. As well as rendering the gene inaccessible to nucleases and transcription factors, this structure may also interfere with the initiation of replication. The cell may be able to distinguish between genes which are required to be in a transcriptionally competent state and genes whose expression will never be required. Such discrimination may be achieved by cytosine methylation (31). Genes which are not required are methylated and assume a condensed chromatin configuration which renders them late replicating. The pattern of methylation is inherited by means of a maintenance methylase and generally can only be altered by interference with the action of this enzyme at replication (although certain transcription factors may be able to override the inhibitory action of methylation). This theme is considered further in Chapter 6.

9. Amplification

Although it is important that each replicon initiates only once per cell cycle, control over this event is not complete. The frequency, in normal cells, of a second initiation occurring in the same replicon within one S-phase is normally very low (32), but for some genes (e.g. *dhfr*), especially in transformed cells, it can be as high as one in a thousand and this frequency can be increased more than tenfold by treatment with inhibitors of DNA synthesis (e.g. hydroxyurea, methotrexate) or DNA damaging reagents (e.g. carcinogens and UV light) (33). Schimke has proposed that treatment with inhibitors traps cells at the G1/S-phase boundary which is the time window in which certain genes (such as *dhfr* and others whose products may regulate initiation) are transcribed. Accumulation of these proteins would precipitate a second initiation at those origins which have just initiated for the first time. Thus the vulnerability to overreplication may depend on the time of replication of a gene within S-phase (Section 8) and the genetic background of the cell. Extensive amplification of particular genes can be obtained by selection for overproduction of the gene product and this is particularly pertinent when the inducing agent is also the generating agent and the selective agent as is the case with methotrexate selection for amplification of the *dhfr* gene.

Selection is a gradual process. The first stage requires the incubation of cells in a low level of inhibitor (e.g. methotrexate) which selects for those cells which have undergone the primary amplification event which is associated with DNA rearrangement and the formation of novel junctions (34). This probably brings the selected gene into a position adjacent to regions of the chromosome which favour further amplification. Incubation in successively higher concentrations of inhibitor selects for those cells which are capable of higher and higher levels of amplification, but further primary rearrangements do not occur at this time.

The actual mechanism of reinitiation is not known and several models have been presented, each of which is particularly suitable in explaining one set of findings (34). A favoured model is that of *onion-skin replication* (*Figure 4.5*), for which evidence has been obtained in the natural amplification of the chorion genes of *Drosophila* (see below). This resembles the situation found in the *E. coli* replicon in rapidly growing cells (*Figure 3.2*). Recombination across a replicating replicon can lead to *double rolling circle replication* which is believed to be the mechanism used for amplification of the copy number of the yeast plasmid (Chapter 3, Section 4.5). *Sister chromatid exchange* mechanisms known to be

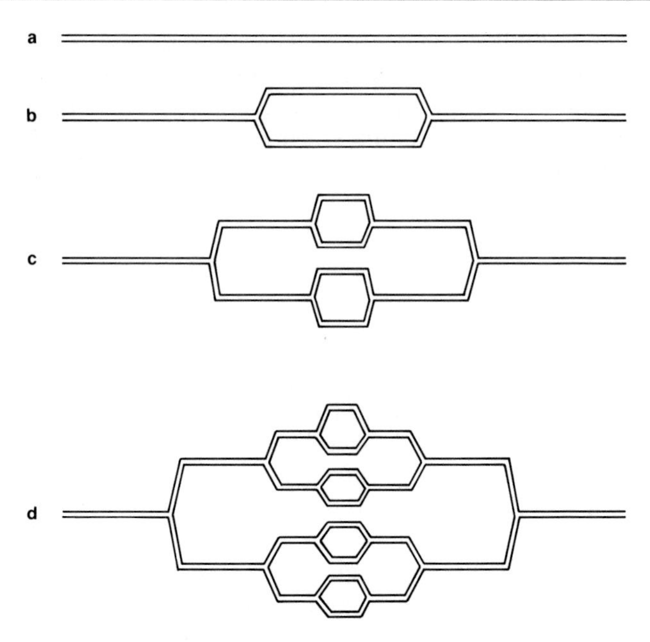

Figure 4.5. Onion skin replication is the multiple reinitiation of replication from a particular origin several times within one S-phase. It results in a nested series of genes which are an ideal substrate for recombination reactions. These could lead to excision of cyclic minichromosomes (double minutes) or the formation of long tandem repeats.

involved in increasing the number of rRNA genes in *Drosophila* are also likely to be involved in the amplification of large stretches of DNA.

The consequences of overreplication of a region of a chromosome depend on the type of recombination event which follows. The chromosomal aberrations found include fragmentation, inversion, sister chromatid exchange, the generation of extrachromosomal elements (double minutes or DMs), and endoreduplication which can result in two or more copies of the replicon becoming established as a tandem repeat. In some cells in which the *dhfr* gene has been amplified many-fold, multiple tandem repeats are readily visible as homogeneously-staining regions (HSRs—also known as extended chromosomal regions or ECRs) in elongated chromosomes. HSRs are relatively stable as they are associated with the chromosome's normal segregation apparatus but double minutes lack centromeres and tend to be rapidly lost from the cell when the selection pressure is removed.

There are some natural processes where multiple tandem repeats are generated within one S-phase at a specific time during development (e.g. the rRNA genes of *Xenopus* and *Drosophila* and the chorion genes of *Drosophila*) (35). An extreme example is found in the vegetative cells of many lower eukaryotes (e.g. *Tetrahymena* and *Oxytricha*) where the required genes from the germ-line, micronuclear DNA are excised and amplified many-fold to form the macronuclear DNA. The *Tetrahymena* rRNA gene is duplicated following excision to form an inverted repeat, both ends of which carry telomeric sequences (Chapter 5, Section 4) and it is this dimer that undergoes further amplification by multiple replication from an internal origin (see Section 3).

10. Conclusions

Chapter 3 described the evidence that there are unique origins of replication in single replicon systems and this theme has been reinforced in this chapter. We are, however, far from any firm conclusions and one recent paper (36) suggests that, at least for the amplified *dhfr* gene, multiple initiations might occur over quite a broad zone (that is, over a 28 kb region) whereas, another paper (38) locates the origin to a 450 bp region. Although mechanisms have been indicated whereby initiation is brought about, relatively little has been said about the control of initiation. It is important to initiate once, and only once, per cell cycle; and it is not easy to see how such control can be exerted particularly in higher eukaryotic systems where multiple origins are present, each initiating at a characteristic time in S-phase. The actual timing of initiation of a replicon may depend on the affinity of an origin for a matrix attachment site, but the regulation of initiation requires the periodic availability of initiation factors. Such factors may lack a nuclear localization signal and hence be restricted to the cytoplasm except during mitosis when the nuclear envelope is transiently broken down (37). It would be at this time when these activating factors would bind to the origins of replication (or to the matrix attachment sites). Excess nuclear factor would be removed in G1-phase and the bound factors would be used up at initiation of replication, thereby preventing further initiation until new factors could gain access to the nucleus at mitosis.

11. Further reading

Adams, R.L.P. and Burdon,R.H. (1986) *The Molecular Biology of DNA Methylation*, Springer Verlag, New York.

DePamphilis,M.L. (1988) Transcriptional Elements as Components of Eukaryotic Origins of DNA Replication. *Cell*, **52**, 635.

Eckdahl,T.T. and Anderson,J.N. (1990) Conserved DNA Structures in Origins of Replication. *Nucl. Acids Res.*, **18**, 1609.

Harland,R. (1981) Initiation of DNA Replication in Eukaryotic Chromosomes. *Trends in Biochem. Sci.*, **6**, 71.

12. References

1. Zakian,V.A. (1976) *J. Mol. Biol.*, **108**, 305.
2. Hand,R. (1978) *Cell*, **15**, 317.
3. Heintz,N.H. and Hamlin,J.L. (1982) *Proc. Natl. Acad. Sci. USA*, **79**, 4083.
4. Linskens,M.H.K. and Huberman,J.A. (1988) *Mol. Cell. Biol.*, **8**, 4927.
5. Brewer,B.J. and Fangman,W.L. (1988) *Cell*, **55**, 637.
6. Bedinger,P., Hochstrasser,M., Jongeneel,C.V. and Alberts,B.M. (1983) *Cell*, **34**, 115.

7. Trempe,J.P., Lindstrom,Y.I. and Leffak,M. (1988) *Mol. Cell. Biol.*, **8**, 1957.
8. Leffak,M. and James,C.D. (1989) *Mol. Cell. Biol.*, **9**, 586.
9. Campbell,J. (1988) *Trends in Biochem. Sci.*, **13**, 212.
10. Kipling,D. and Kearsey,S.E. (1990) *Mol. Cell. Biol.*, **10**, 265.
11. McWhinney,C. and Leffak,M. (1990) *Nucl. Acids Res.*, **18**, 1233.
12. Krysan,P.J., Haase,S.B. and Calos,M.P. (1989) *Mol. Cell. Biol.*, **9**, 1026.
13. Brewer,B.J. and Fangman,W.L. (1987) *Cell*, **51**, 463.
14. Huberman,J.A., Spotila,L.D., Nawotka,K.A., El-Assouli,S.M. and Davis,L.R. (1987) *Cell*, **51**, 473.
15. Umek,R.M., Linskens,M.H.K., Kowalski,D. and Huberman,J.A. (1989) *Biochim. Biophys. Acta*, **1007**, 1.
16. Handeli,S., Klar,A., Meuth,M. and Cedar,H. (1989) *Cell*, **57**, 909.
17. Vassilev,L. and Johnson,E.M. (1988) *Nucl. Acids Res.*, **127**, 7693.
18. Russev,G. and Vassilev,L. (1982) *J. Mol. Biol.*, **161**, 77.
19. Brown,E.H., Iqbal,M.A., Stuart,S., Hatton,K.S., Valinsky,J. and Schildkraut,C.T. (1987) *Mol. Cell. Biol.*, **7**, 450.
20. Graham,C.F. (1966) *J. Cell Sci.*, **1**, 363.
21. Buongiorno-Nardelli,M., Mikcheli,G., Carri,M.T. and Marilley,M. (1982) *Nature (London)*, **298**, 100.
22. Jackson,D.A. and Cook,P.R. (1986) *EMBO J.*, **5**, 1403.
23. Umek,R.M. and Kowalski,D. (1988) *Cell*, **52**, 559.
24. Caddle,M.S., Lussier,R.H. and Heintz,N.H. (1990) *J. Mol. Biol.*, **211**, 19.
25. Goldman,M.A., Holmquist,G.P., Gray,M.C., Caston,L.A. and Nag,A. (1984) *Science*, **224**, 686.
26. Dhar,V., Skoultchi,A.I. and Schildkraut,C.L. (1989) *Mol. Cell. Biol.*, **9**, 3524.
27. Enver,T., Brewer,A.C. and Patient,R.K. (1988) *Mol. Cell. Biol.*, **8**, 1301.
28. Jones,K.A., Kadonaga,J.T., Rosenfeld,P.J., Kelly,T.J. and Tjian,R. (1987) *Cell*, **48**, 79.
29. Cheng,L. and Kelly,T.J. (1989) *Cell*, **59**, 541.
30. Pruijn,G.J.M., van Driel,W. and van dar Vliet,P.C. (1986) *Nature (London)*, **322**, 656.
31. Adams,R.L.P. (1990) *Biochem. J.*, **265**, 309.
32. Wright,J.A., Smith,H.S., Watt,F.M., Hancock,M.C., Hudson,D.L. and Stark,G.R. (1990) *Proc. Natl. Acad. Sci. USA*, **87**, 1791.
33. Schimke,R.T., Sherwood,S.W., Hill,A.B. and Johnston,R.N. (1986) *Proc. Natl. Acad. Sci. USA*, **83**, 2157.
34. Stark,G.R., Debatisse,M., Giulotto,E. and Wahl,G.M. (1989) *Cell*, **57**, 901.
35. Kafatos,F.C., Orr,W. and Delidakis,C. (1985) *Trends in Genet.*, **1**, 301.
36. Vaughn,J.P., Dijkwel,P.A. and Hamlin,J.L. (1990) *Cell*, **61**, 1075.
37. Blow,J. and Laskey,R.A. (1989) *Nature (London)*, **332**, 546.
38. Burhans,W.C., Vassilev,L.T., Caddle,M.S., Heintz,N.H. and DePamphilis,M.L. (1990) *Cell*, **62**, 955.

5

Termination of replication

1. Introduction

The fact that DNA polymerase requires a primer gives rise to a major problem when it comes to the replication of the very ends of a linear chromosome (*Figure 5.1*). Even if the primase initiates at the extreme 3′ end of the template DNA, a few nucleotides will remain single-stranded when the primer is removed. One of the strategies (Section 3) to circumvent this problem is to have circular chromosomes, but this creates its own difficulties when the two replication forks converge on one another. Similar difficulties arise in the chromosomes of eukaryotes when adjacent replicons fuse, and these problems are discussed in Section 2.

The point has already been made (Chapter 4, Section 3) that an advancing transcription complex can inhibit the progression of the replication fork, and special mechanisms are present that ensure such clashes are kept to a minimum (Section 2).

Another major problem with the ends of linear duplex DNA molecules is that they are highly recombinogenic. Linear fragments of DNA introduced into cells

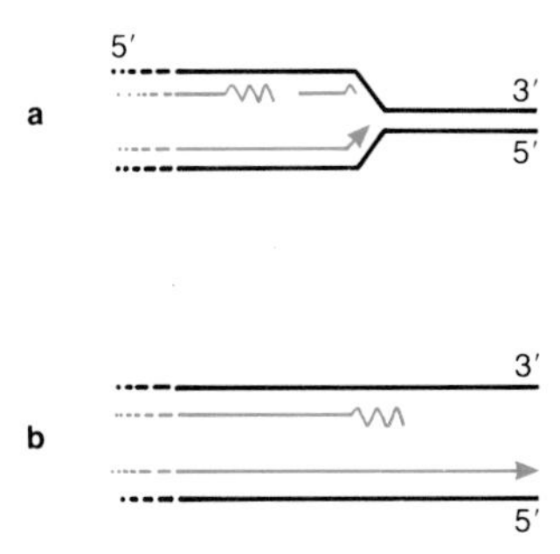

Figure 5.1. Replicating the end. A non-replicated region must remain at the 3′ end of every linear duplex.

are unstable and quickly become integrated into the host chromosomes. It is imperative, therefore, that ends of linear chromosomes should be protected in some way, and some solutions to this problem are considered in Sections 3 and 4.

2. Fusion of replicons

2.1 Unlinking of parental DNA

Figure 5.2 illustrates the situation which arises when two adjacent replicons fuse in eukaryotic chromosomes. A similar situation exists in cyclic duplex molecules as the two replicating forks, diverging from the origin, converge on one another and also, when replication is unidirectional, when one round of replication nears completion.

Helicase action unwinds the parental duplex a little way ahead of the point of action of DNA polymerase, and topoisomerase I can normally act as a swivel to reduce the linking number of the parental duplex. However, as two forks approach one another, topoisomerase I is excluded creating a situation in which

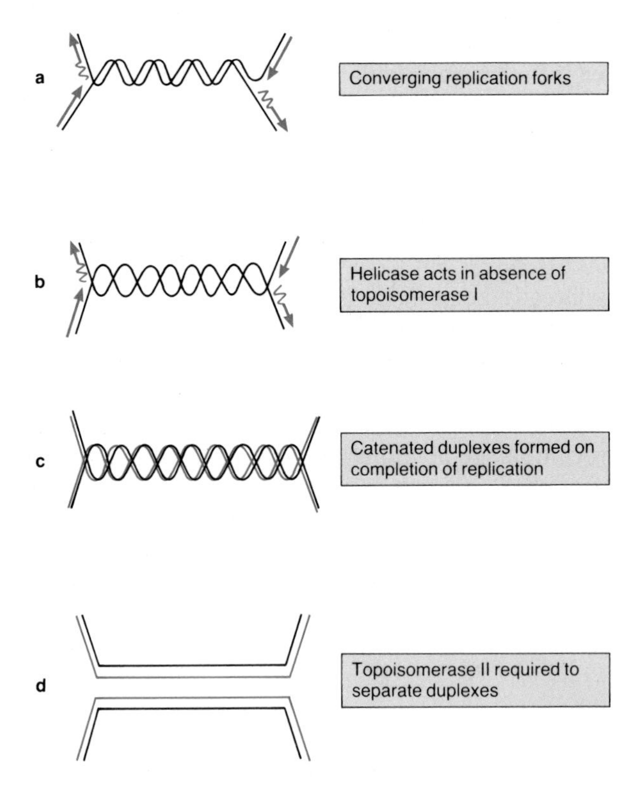
a
Converging replication forks
b
Helicase acts in absence of topoisomerase I
c
Catenated duplexes formed on completion of replication
d
Topoisomerase II required to separate duplexes

Figure 5.2. Replicon fusion. As two replication forks converge **a**, topoisomerase I is excluded **b** and this results in the production of two highly catenated molecules **c** which require topoisomerase II action for resolution **d**.

the two strands of the parental duplex have been separated, but not yet replicated, and are still wound around each other (linked) by up to 20 times (1). This leads to a delay in the completion of replication and almost completed molecules of SV40 DNA accumulate. Accumulation is enhanced in the presence of inhibitors of topoisomerase II (Chapter 2, Section 4) or high concentrations of NaCl (2, 3).

This shows that there is an absolute requirement for topoisomerase II to decatenate the two daughter duplexes of cyclic DNA duplex molecules and also to allow the two chromosomes to separate following replication in eukaryotes. Topoisomerase II is found typically associated with the matrix which is the site of termination of replication (Chapter 4, Section 7) (4) and inhibitors of topoisomerase II prevent the normal separation of chromosomes at mitosis and meiosis.

2.2 Termination signals

The foregoing observations indicate that termination takes place at the point where replication forks converge, and that this is at a particular position. Assuming a constant rate of fork movement, such a point would tend to be midway between initiation sites (or opposite the initiation site in a cyclic chromosome). However, as pointed out in Chapter 4, Section 3, this is not always the case as replication forks can be stalled when they encounter an oncoming transcription complex. In many cases (e.g. in SV40 DNA) stalling will occur only if one fork moves faster than the other as the point opposite the origin is where both clockwise and counterclockwise transcription terminate.

With larger replicons, such as those of *E. coli* and *B. subtilis* and plasmids RK6 and R100, it is insufficient to rely on each arm of the fork moving around the chromosome at the same rate. Specific signals are present to arrest the replication forks as they approach the terminus (5, 6).

The consensus sequence for termination of replication (*ter*) is

AATTAGTATGTTGTAACTAAANT.

This is non-palindromic and blocks only those replication forks approaching from the 5′ end. Thus, a pair of sequences is required to block forks approaching from opposite directions. The sequences are arranged so that a fork passes through the first *ter* sequence and becomes arrested at the second, that is, the two forks enter the region between the *ter* sequences and become trapped there. In *E. coli*, four *ter* sequences are present in the terminus region (*Figure 5.3*). A pair of adjacent sequences (*ter*C and *ter*B) block the fork approaching from the left and a second pair (*ter*D and *ter*A) block the fork approaching from the right. In *E. coli* there is a region of 275 kb between the two pairs of sequences (in which termination must take place) while there are only 73 bp between the *ter* sites in plasmid RK6.

The *ter* sequence binds a protein (*ter*-binding protein, TBP, Tus, or Tau) and the promoter of the *tus* gene is in the *ter*B site indicating autoregulation of Tus expression (7, 8). The presence of *ter* alone has no effect, but binding of Tus to *ter* blocks helicase action. DnaB helicase (translocates 5′ to 3′) and helicase II and Rep (translocate 3′ to 5′) are blocked only when *ter* is oriented such that

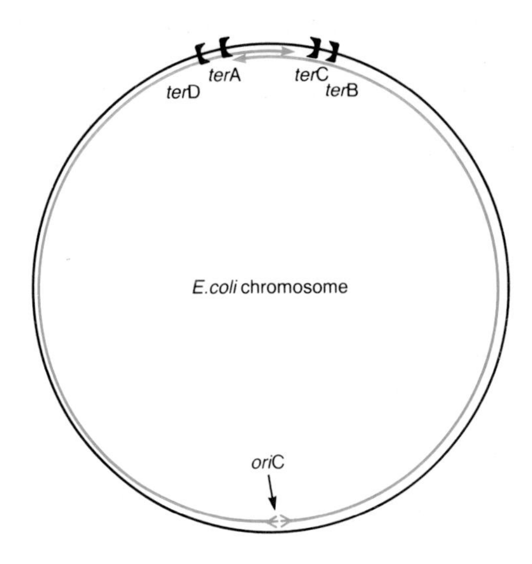

Figure 5.3. Termination sequences in *E. coli*. There are four *ter* sequences, roughly opposite the origin of the *E. coli* chromosome. *ter*D and *ter*A prevent passage of the anticlockwise fork and *ter*C and *ter*B of the clockwise fork, thereby trapping the forks in the termination region.

the helicase approaches from the 5′ end. The replication forks appear to stall as they pass Tus bound to *ter*. In this way, if one fork moves faster than the other, it will be trapped in the terminus region to await the arrival of the counter-moving fork. Such a defined terminus would allow other signals, whose replication might activate cell division, to be located at this point.

3. Termination strategies for linear chromosomes

As indicated in *Figure 5.1* linear chromosomes encounter problems in completion of replication. Some, like the chromosome of phage lambda have 12- or 19-nucleotide long, 'sticky' ends which can anneal, and this causes a circular chromosome to form on entry into the host cell. Phage lambda DNA replicates as a θ structure in the early stages of infection (see Chapter 3, Section 4.2) but later replication is by a rolling circle (σ) mechanism which generates long, concatameric, linear molecules. The origin for rolling circle replication is the same as that for early replication and it has been suggested that some θ molecules may be replicating unidirectionally and, rather than terminate after one round of replication, the leading strand may simply displace the tail of the previously synthesized strand to generate the rolling circle. Unit length, linear molecules with cohesive ends are regenerated during packaging by means of lambda terminase (the product of genes *A* and *Nu1*) and host factors. The concatamer is cleaved by terminase at the *cos* site to generate cohesive ends known as right and left. Terminase-bound left ends are inserted into the capsid shell and packaging continues in an ATP dependent reaction until the next *cos* site is encountered when cleavage occurs again (9).

Others, such as the chromosomes of adenovirus or phage ϕ29 have adopted a special mechanism of initiation which circumvents the problem of termination (Chapter 3, Section 6). The single-stranded chromosomes of the parvoviruses (10) and the double-stranded chromosomes of phage such as T7 have adopted an approach involving a specific endonuclease and the details of the latter will now be considered.

The two, terminal 260 bp of phage T7 DNA are direct repeats of one another which enables the replicated molecules (**b** in *Figure 5.1*) to anneal with one another to form dimeric or multimeric molecules (*Figure 5.4*). DNA ligase can seal the gap to produce a linear concatamer containing two or more copies of the phage DNA held together by single copies of the redundant sequence. A mechanism of concatamer resolution was proposed (11, 12), that involved a specific endonuclease nicking both strands of the DNA at the 3′ end of the redundant sequence (*Figure 5.4*) to generate two 3′ OH groups that could serve as primers for T7 DNA polymerase. In a strand displacement reaction, the polymerase would complete both daughter duplexes which would separate as the forks passed one another (*Figure 5.4*).

However, this explanation fits only partially with some of the observations made more recently. Although the T7 gene 3 endonuclease is required for processing concatamers, it alone cannot act in a site specific manner. Furthermore, processing of long concatamers is dependent on the packaging of the DNA and, in the absence of either gene 2 protein (an inhibitor of *E. coli* RNA polymerase) or gene 6 exonuclease, aberrant breakdown of concatamers occurs. Phage T7 gene 18 protein, in combination with a host factor, binds to the redundant sequence and delivers the concatameric sequence to the capsid where processing and packaging occur (cf. phage lambda DNA packaging).

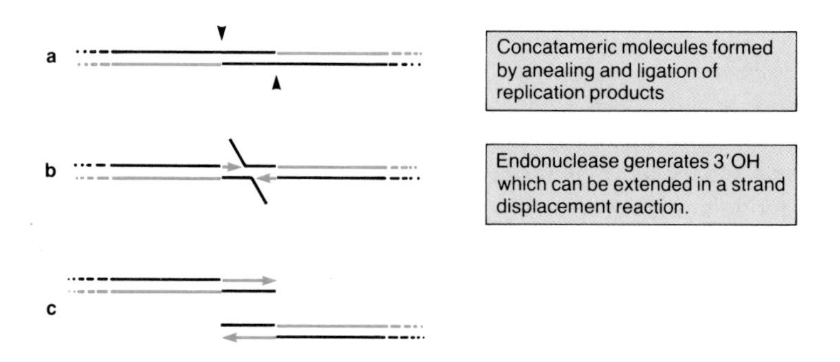

Figure 5.4. Termination of T7 DNA replication—A. One proposal (12) for termination involves using the redundant terminal sequences in the formation of a linear concatamer by annealing and ligating the replication products (see *Figure 5.1*) **a**. Resolution requires a specific endonuclease to generate 3′ OHs which can be extended in a strand displacement reaction **b**,**c**.

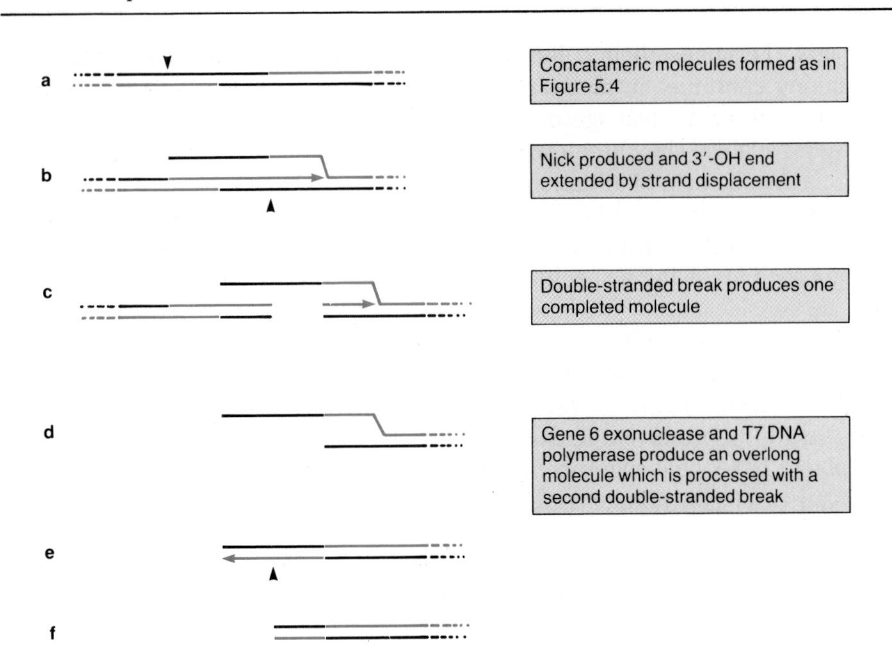

Figure 5.5. Termination of T7 DNA replication—B. An alternative proposal (13) involves a different means of resolution of the catenated molecule **a**. A nick is produced on the parental strand 5′ to the redundant region and the 3′ OH extended by strand displacement **b**. A double-stranded break now produces one complete molecule **c**. Gene 6 exonuclease removes a stretch of redundant DNA **d** and the T7 polymerase produces an overlong duplex **e** which is trimmed with a second double-stranded break **f**.

Figure 5.5 describes a model in which the left- and right-hand ends are produced by distinct mechanisms which is more in keeping with *in vitro* results (13).

4. Telomeres

Short, linear chromosomes present in single-celled eukaryotes such as *Tetrahymena*, *Oxytricha*, trypanosomes and yeast (see Chapter 4, Section 9) have provided models that have led to a partial understanding of the structure and replication of the telomeres (ends) of the chromosomes of higher eukaryotes.

Restriction fragments introduced into animal cells rapidly integrate into the host chromosomes indicating that the cut ends are highly susceptible to recombination. This is clearly different from the natural ends of chromosomes. When telomeres are attached to the ends of linear DNA they can protect that DNA from recombination and this is important in the construction of yeast artificial chromosomes. It may be the actual structure of the telomeric DNA that protects it from recombination, or it may be the result of a sequence-specific interaction with a protein that stabilizes the end of the DNA (14).

The telomere also provides a structure that allows the replication of the very end of a chromosome. Chromosomes have been found in some *Drosophila* mutants that have lost their telomeres (and their ability to repair double-stranded breaks in their DNA) and these chromosomes get shorter by about 75 bp per generation (15).

What distinguishes telomeres from the ends of DNA at the site of restriction enzyme cleavage is that, in the former, the one strand of DNA overhangs and folds back to hydrogen bond with itself in a two- or four-stranded structure. Thus, at the very end the two strands are covalently linked and there is a subterminal nick in one strand (*Figure 5.6*).

Telomeres are also characterized by the presence of multiple repeats of very short sequences such as TTGGGG (*Tetrahymena*) or TTAGGG (human). It is always the G-rich strand that is the longer, such that, when it folds back on itself, it can be stabilized by quartets of Gs, in the *syn* and *anti* configurations, linked together by hydrogen bonds (*Figure 5.7*) (16, 17).

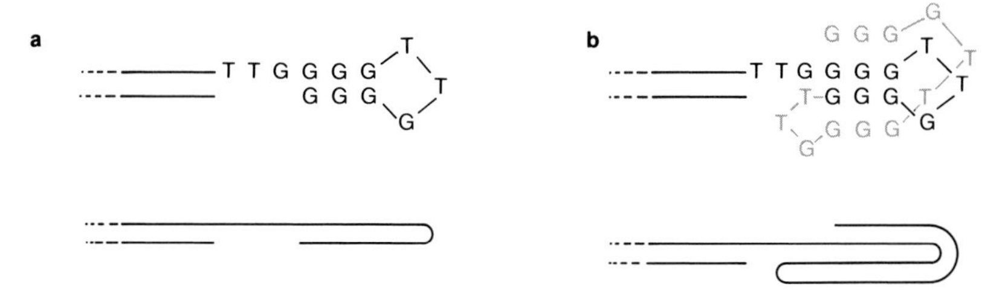

Figure 5.6. Telomere structure. Two ways in which the G-rich strand could fold back on itself to form a two **a** or a four **b** stranded structure.

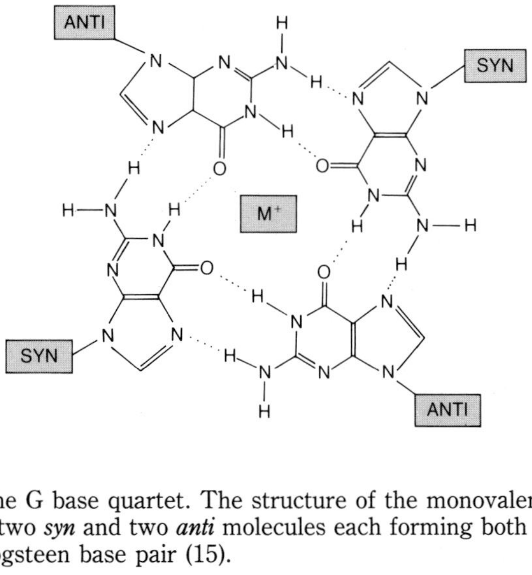

Figure 5.7. The G base quartet. The structure of the monovalent ion stabilized G quartet involving two *syn* and two *anti* molecules each forming both the donor and the acceptor of a Hoogsteen base pair (15).

The actual number of repeats at the end of a chromosome is not constant but increases during vegetative growth only to return to a low number as cells enter the stationary phase. In *Tetrahymena* growth in the number of repeats is brought about by the action of an enzyme known as a telomerase. This ribonucleoprotein adds nucleotides on to the G-rich strand, one at a time (18). The order of addition is dictated by a short region in the RNA of the enzyme acting as a template (*Figure 5.8*). Thus, when a *Tetrahymena* chromosome is introduced into yeast it is given a yeast telomeric sequence (19).

Telomerase thus acts to extend the 3′ end of the chromosome, such that failure of the C-rich strand to initiate at the very end is rendered unimportant. It is presumed that synthesis of the C-rich strand is initiated using a normal RNA primer, though the possibility does exist that the G-rich strand could act as a primer after folding back.

Apart from its being a ribonucleoprotein, telomerase is of interest in that it is a reverse transcriptase present in most, if not all, eukaryotic cells. As well

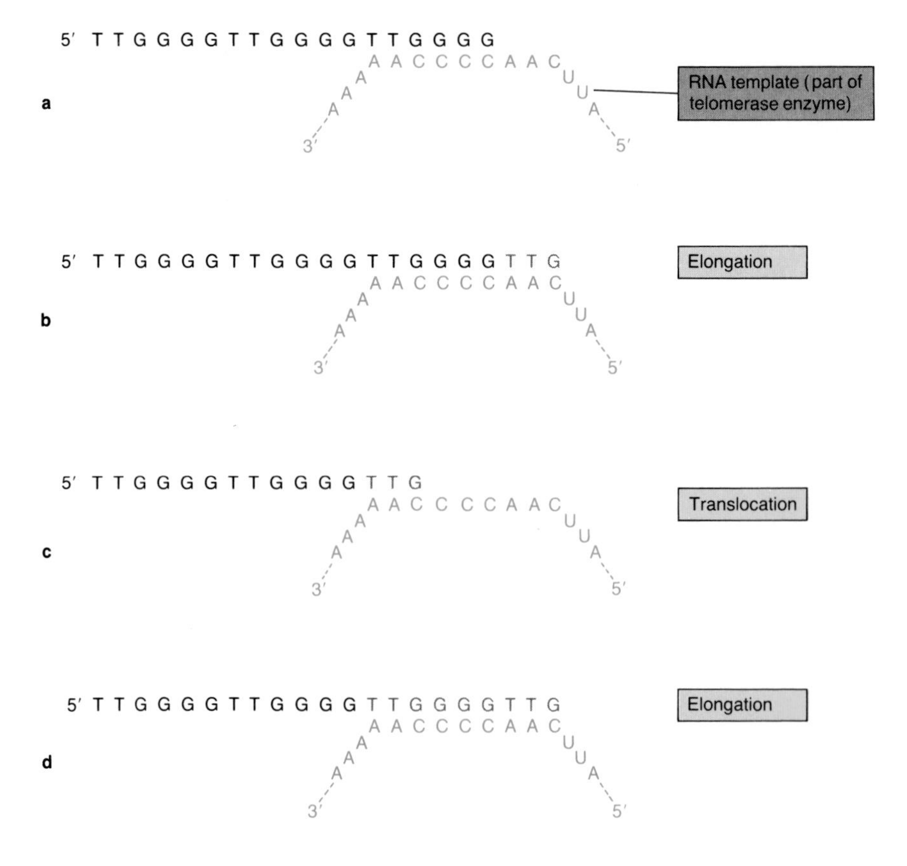

Figure 5.8. Telomerase action. The *Tetrahymena* telomerase is a ribozyme containing and RNA template (orange) upon which the 3′ end is extended using alternative phases of elongation, **b** and **d** and translocation **c**.

as acting at the ends of chromosomes, it could act at internal break points to convert them into telomeres; or on a displaced daughter strand at the replication fork which would lead to the production of internal telomeric sequences.

This model is only partly able to explain the synthesis of telomeres in yeast where non-reciprocal recombination events have been observed to occur during telomere replication. In contrast to the situation in ciliates, the telomeric repeats in *Saccharomyces*, *Dictyostelium*, and *Schizosaccharomyces* are more variable (i.e. $C_{1-3}A$, $C_{1-8}T$, and $C_{1-6}G_{0-1}GTA_{1-2}$ respectively) and so cannot be synthesized by a simple telomerase as described above, although a family of telomerases has not been ruled out. An alternative explanation of these findings, is that the telomeric, single-stranded extension can invade a region of double-stranded telomeric DNA and acquire a ready made telomere by recombination (*Figure 5.9*) (20). Indeed some 'telomeric' sequences occur at internal sites and, in vertebrates, these are sites of chromosome fragility and recombination (21). The five chromosomes of *Drosophila* do not appear to have these simple telomeric sequences but, rather, are characterized by long, complex telomeres containing copies of HeT repetitive DNA (14). The broken ends of *Drosophila* chromosomes referred to above can be 'healed' by reacquiring HeT sequences.

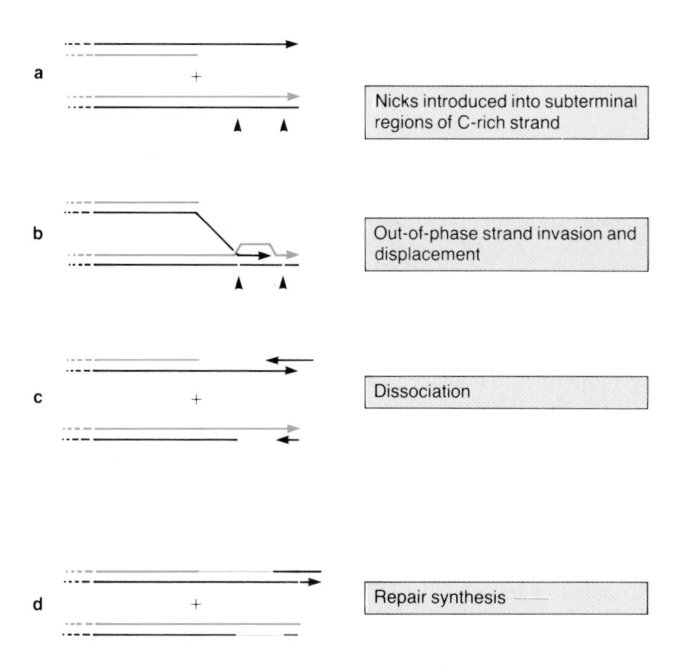

Figure 5.9. Formation of telomeres by recombination. Nicks introduced into the C-rich strand might facilitate out-of-phase strand invasion by the single-stranded 3′ extension with displacement of the newly synthesized 3′ end of the sister chromatid **b**. On dissociation, a 5′ fragment could be transferred **c** and repair synthesis (grey) would generate one extended and one complete duplex molecule **d** (18).

5. Conclusion

In Section 2 of Chapter 1, the suggestion was made that DNA polymerase requires a primer in order to exclude undesirable initiations. It is clear from this chapter that such a requirement leads to severe problems when it comes to completion of replication. A number of different strategies are used to circumvent these problems, and it is probable that these all evolved from the use of specific terminal structures (redundant sequences). In many cases this has led to the adoption of a cyclic chromosome and the chromosomes of phage lambda and adenovirus may represent intermediate stages.

6. Further reading

Brewer,B.J. (1988) When Polymerases Collide: Replication and the Transcriptional Organisation of the *E. coli* Chromosome. *Cell*, **53**, 679.
Kuempel,P.L., Pelletier,A.J. and Hill,T.M. (1989) Tus and the Terminators: The Arrest of Replication in Prokaryotes. *Cell*, **59**, 581.
Tapper,D.P. and De Pamphilis,M.L. (1980) Preferred DNA sites are involved in the Arrest and Initiation of DNA Synthesis during Replication of SV40 DNA. *Cell*, **22**, 97.
Zakian,V.A., Runge,K. and Wang,S.-S. (1989) How does the End Begin? *Trends in Genet.*, **6**, 12.

7. References

1. Snapke,R.M., Powelson,M.A. and Strayer,J.M. (1988) *Mol. Cell. Biol.*, **8**, 515.
2. Richter,A. and Strausfeld,U. (1988) *Nucl. Acids Res.*, **16**, 10119.
3. Marians,K.J. (1987) *J. Biol. Chem.*, **262**, 10362.
4. Adachi,Y., Kas,E. and Laemmli,U.K. (1989) *EMBO J.*, **8**, 3997.
5. Hill,T.M., Pelletier,A.J., Tecklenburg,M.L. and Keumpel,P.L. (1988) *Cell*, **55**, 459.
6. Hidaka,M., Akiyama,M. and Horiuchi,T. (1988) *Cell*, **55**, 467.
7. Khatri,G.S., MacAllister,T., Sista,P.R. and Bastia,D. (1989) *Cell*, **59**, 667.
8. Lee,E.H., Kornberg,A., Hidaka,M., Kobayashi,T. and Horiuchi,T. (1989) *Proc. Natl. Acad. Sci. USA*, **86**, 9104.
9. Feiss,M. and Becker,A. (1983) In *Lambda II*, 305. Hendrix,R.W., Roberts,J.W., Stahl,F.W. and Weisberg,R.A. (eds). Cold Spring Harbor Press.
10. Chen,K.C., Tyson,J.J., Lederman,M., Stout,E.R. and Bates,R.C. (1989) *J. Mol. Biol.*, **208**, 283.
11. Kelly,T.J. and Thomas,C.A. (1969) *J. Mol. Biol.*, **44**, 459.
12. Watson,J.D. (1972) *Nat. New Biol.*, **239**, 197.
13. White,J.H. and Richardson,C.C. (1987) *J. Biol. Chem.*, **262**, 8851.
14. Price,C.M. (1990) *Mol. Cell. Biol.*, **10**, 3421.
15. Biessmann,H., Mason,J.M., Ferry,K., d'Hulst,M., Valgeirsdottir,K., Traverse,K.L. and Pardue,M.-L. (1990) *Cell*, **61**, 663.
16. Henderson,E.R. and Blackburn,E.H. (1989) *Mol. Cell. Biol.*, **9**, 345.
17. Williamson,J.R., Raghuraman,M.K. and Cech,T.R. (1989) *Cell*, **59**, 871.
18. Morin,G.B. (1989) *Cell*, *59*, 521.
19. Shampay,J., Szostak,J.W. and Blackburn,E.H. (1984) *Nature (London)*, **310**, 154.
20. Pluta,A.F. and Zakian,V.A. (1989) *Nature (London)*, **337**, 429.
21. Hastie,N.D. and Allshire,R.C. (1989) *Trends in Genet.*, **5**, 326.

6

DNA packaging

1. Introduction

Little is known of the packaging of DNA in bacteria, though the chromosome is looped in a manner similar to the eukaryotic chromosome (Chapter 4, Section 7) and protein HU may be involved in folding the DNA. In viruses the DNA is finally packaged into the virion and, already, some mention of this has been made with respect to ϕX174 (Chapter 3, Section 3.1) and phage lambda (Chapter 5, Section 3).

This chapter will concentrate primarily on the nucleus of eukaryotic cells where the DNA is packaged with small, basic proteins (histones) into chromatin. Eight histone molecules (two each of histones H2A, H2B, H3 and H4, together forming the histone octamer) associate together to form a nucleosome core, around which is wound the DNA. 146 bp of DNA are closely associated with the core with a variable number, 10–80 bp, forming a linker joining adjacent nucleosomes. A fifth histone, the very lysine-rich H1, is associated with the linker DNA in transcriptionally inactive chromatin and H1 crosslinks help to condense the DNA even further to form heterochromatin.

The arginine-rich histones, H3 and H4, are highly conserved and form the central part of the nucleosome, while the lysine-rich H2A and H2B are associated with the sides of the disc-like nucleosome. The 146 bp of core DNA are wound around this structure 1.75 times, and the constraints this places on DNA structure mean that nucleosomes are more stable when formed in regions of DNA which are more easily bent.

The synthesis of histones is largely restricted to S-phase and there is a link between the synthesis of the two which is considered in Section 2.

There has been considerable interest over many years as to how the machinery of transcription and replication is affected by the presence of the chromatin proteins, and whether or not the nucleosomal structure is displaced or dissociated in order to allow the machinery to pass. Obviously, as the histones are associated with both strands of the parental duplex DNA, this association cannot remain when the two strands part during replication, and the fate of parental histones and the location of new histones is considered in Section 3.

In addition to the histones, there are other proteins associated with the DNA. These include the transcription factors, and their disposition at replication is crucial to the pattern of gene expression in the daughter cells. This is one aspect of epigenetic information which is considered in Section 4 along with post-synthetic, covalent modification of the DNA.

2. Histone synthesis and nucleosome assembly

There are developmentally-regulated histone genes, for instance in the sea urchin. Such genes, present in very high numbers, are active in the early cleavage stages when large amounts of variant histones are required to package the rapidly replicating DNA. In most organisms, however, there are multiple copies of the genes for most of the histones and, in general, these appear to fall into two classes.

Expression of the majority of histone genes is linked to replication, such that histone synthesis is tightly coupled with DNA replication. Inhibition of histone synthesis leads to a slowing down of DNA replication which eventually stops after about 30 minutes. Similarly, inhibition of DNA replication, for instance with hydroxyurea, leads to an almost complete cessation of histone synthesis similar to that which occurs at the end of S-phase (1). This inhibition appears to act partly through a translational control as the half life of the histone mRNA falls dramatically at the end of S-phase. Additionally, there is control over transcription and this may act by a single mechanism leading to the activation of a battery of transcription factors each responsible for the activity of a series of S-phase related genes (2). Histones from this class of genes are the major contributors to the new nucleosomes which associate with the newly replicated DNA (see Section 3). A minority of histone genes are active throughout the cell cycle and histones produced from these genes are able to exchange to a greater or lesser extent with the histones in established nucleosomes.

The histone composition of nascent nucleosomes has been established by following the incorporation of radioactive, density-labelled amino acids into nucleosomes associated with DNA at various times post replication (3, 4). Within an hour of replication, nucleosomes were composed of all light histones (parental octamers) or all heavy histones (daughter octamers). This conclusion is partly obscured by the fact that, even within a DNA-bound octamer, there is some turnover of histones at times other than S-phase. This turnover is slight for H3 and H4, but is readily detected for H2A and H2B. Bound H1 undergoes frequent exchange with the free pool of H1 molecules (5).

From analyses made at short times after replication, it is clear that histones H3 and H4 very rapidly form an immature nucleosome around which the DNA is able to pack. These structures are formed within 1000 bp from the site of DNA synthesis and assembly requires the presence of a nuclear factor (N1). In the presence of a second factor (nucleoplasmin) the nucleosomes mature by addition of histones H2A and H2B to the faces of the immature nucleosomes, and this happens between 1000 and 10 000 bp from the replication fork (i.e. up to 50 nucleosome diameters). Histone H1 is added only much later (6–8).

This time course does not agree very well with the finding that newly synthesized chromatin is hypersensitive to nucleases for up to 20 minutes post replication (9). This is approximately the time required for the synthesis and joining together of replicons and it has been proposed that large regions of the chromosome undergo replication together, and it is only when a number of adjacent replicons are complete that the chromatin structure returns to normal.

3. Nucleosome segregation

A number of different scenarios have been proposed (10) (*Figure 6.1*) to explain what might happen to parental nucleosomes at replication, and experiments to distinguish between them have seldom been conclusive.

If it is assumed that parental octamers are fairly stable (see Section 2) but that they are displaced from the DNA at replication, they might be expected to associate at random with the two daughter duplexes. Some evidence for this has been obtained from a study of the location of newly synthesized histone octamers, while other experiments appear to show a segregation of parental octamers to the *leading* side of the replication fork.

Convincing evidence for the former was obtained by crosslinking DNA strands *in vivo* with trimethylpsoralen which can act only on the linker DNA. On isolation and denaturation of the DNA, bubbles indicated the position of nucleosomes (*Figure 6.2*) (11). When the experiment was performed in the presence of cycloheximide to prevent synthesis of new histones, it was clear that the parental nucleosomes had segregated at random to the two arms at the fork. More recent

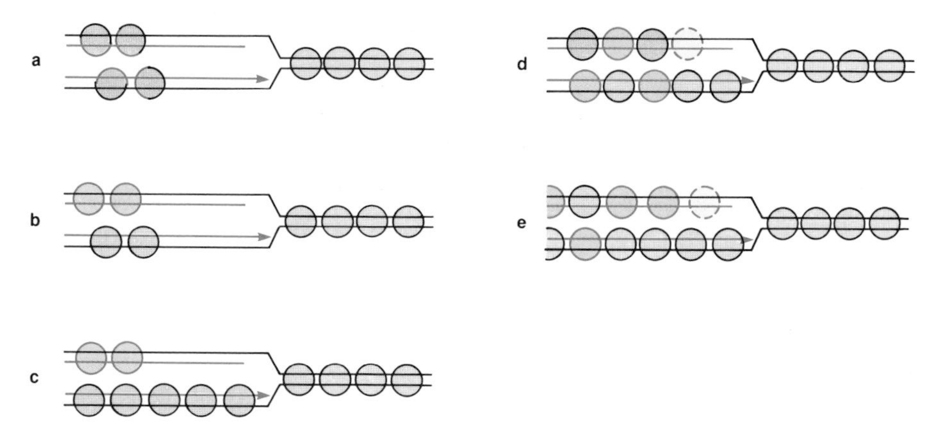

Figure 6.1. Nucleosome segregation. Several different possible segregation patterns are shown, and others could be envisioned. The parental nucleosomes are black and the new ones orange, with the dashed ones being as yet incomplete. The final diagram **e** may be the most correct.

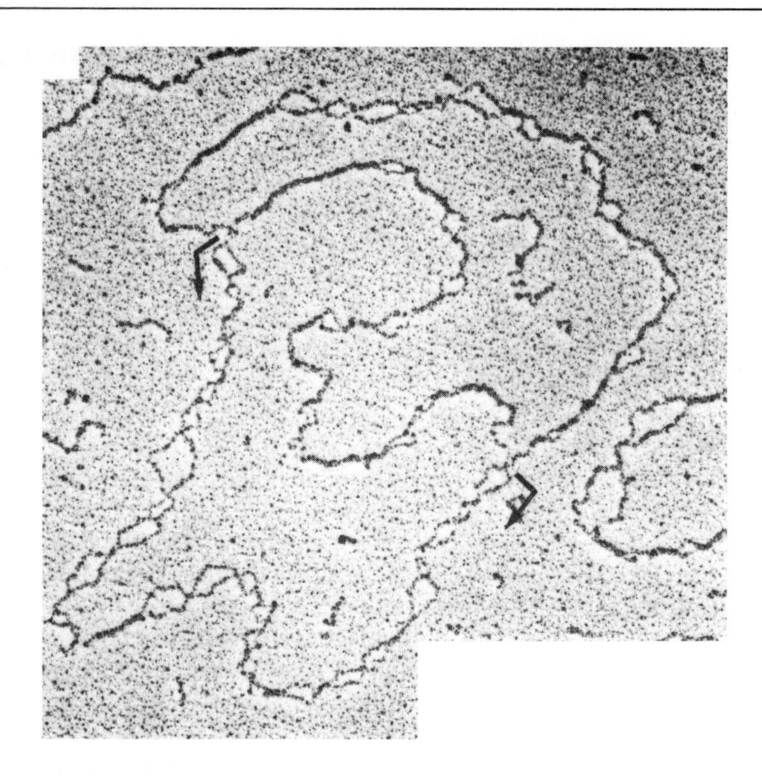

Figure 6.2. Electron micrograph of a crosslinked fork. SV40 minichromosomes were isolated after growth in cycloheximide for 20 minutes and crosslinked with trimethylpsoralen. The DNA was then extracted and prepared for electron microscopy (11). Bubbles represent the location of nucleosomes which are essentially absent on both sides of the replication fork for a distance of about 250 bp. There is apparently random segregation of parental nucleosomes to the two arms of the fork. (With the kind permission of the authors and publishers.)

experiments using a phage T4 replicating complex (12) confirm the segregation of nucleosomes to both sides of the fork but indicate a strong bias towards the leading strand and, in other experiments performed in the presence of cycloheximide, forks are found at which only one of the daughter arms is covered with nucleosomes (13). Again, using the SV40 minichromosome, it was inferred that the parental nucleosomes associated with the coding strand at replication. (In SV40 the coding strand is known to be the leading strand—see Chapter 3, Section 4.4) (14). Acceptance of such conclusions is fundamental to one of the methods of locating origins of replication as discussed in Chapter 4, Section 5.

If nucleosomes are released from the DNA in the vicinity of the replicating fork they might be expected to be the first to reassociate with the daughter duplexes once the replication fork has passed. Nevertheless, no redistribution to competitor DNA is seen (12) and it appears most likely that the replicating complex is, in some way, able to replicate past a bound nucleosome (or at least past a nucleosome bound to the leading strand). Certainly, non-random association requires either that the nucleosomes remain associated with one of the parental strands as the replication fork passes or that they preferentially associate with the leading arm of the fork after replication. The basis for such possible reactions is not known but may simply reflect a lower affinity of the octamer for the single-stranded region of DNA that exists transiently on the lagging side of the fork.

An acceptable model involves a nucleosome jumping or sliding from the parental duplex DNA ahead of the fork to the duplex DNA on the leading arm of the fork. As the nucleosome moves, the duplex in unwound and a new primer is made on the lagging side of the fork. This is consistent with the actual size of the Okazaki pieces in eukaryotes which is much smaller than their size in prokaryotes (Chapter 1, Section 4).

4. Conservation of epigenetic information

As well as information carried in the sequence of the four bases (genetic information), information can be transferred from one generation to the next by so-called epigenetic mechanisms. These include the continued association of specific proteins with both daughter and parental DNA molecules (protein templating) and the ticketing of certain bases by methylation.

4.1 Protein templating

As well as histones, the DNA is associated with a variety of other proteins including transcription factors. The presence of particular transcription factors are essential for the activity of a gene. If these factors are displaced from the DNA at replication, this could lead to a change in the pattern of transcription. Although such changes do occur during development, most dividing cells retain their characteristic spectrum of active and inactive genes. It is assumed that, following replication, there is a competition between the binding of histones and the binding of specific transcription factors (15). However, when present in the solution surrounding the replicating chromosome, the binding of transcription factors takes precedence over the binding of the histone octamers leading to a nucleosome-free region around active promoters (16). Only occasionally has evidence been obtained that transcription factors can disrupt already established nucleosome structure.

Transcriptionally-active chromatin is characterized by a deficiency of histone H1 and an increased sensitivity to nucleases, a sensitivity that is lost at replication. Sensitivity is regained within about 20 nucleosomes (4000 bp) of the fork and it has been proposed that this may depend on a templating effect in which one subunit of a dimeric protein remains bound to each parental DNA strand at replication and thereby determines the location of that protein on the daughter duplex (17,18) (*Figure 6.3*).

On replication of active chromatin (as opposed to bulk chromatin) the parental nucleosomes tend to dissociate, such that the nucleosomes after the fork are made up partly of old and partly of newly synthesized histones (19). It is possible

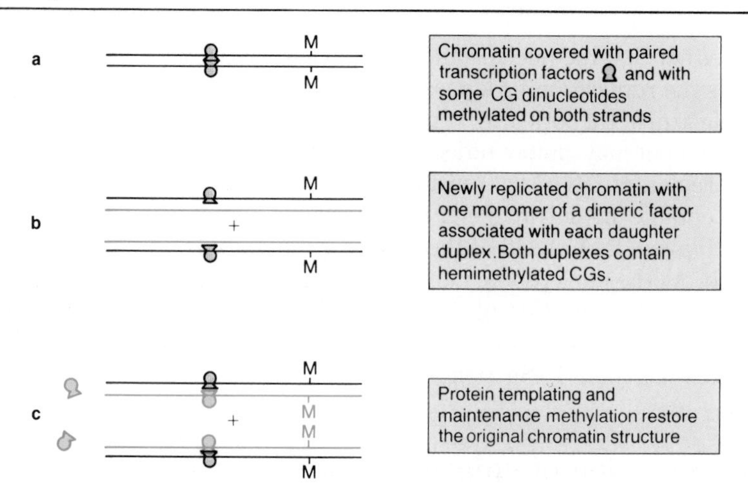

Figure 6.3. Protein templating and maintenance methylation. The top diagram **a** shows chromatin with paired transcription factors and with some CG dinucleotides methylated on both strands. The second diagram **b** shows newly replicated chromatin with one monomer of the dimeric factor associated with each daughter strand. Both duplexes contain hemimethylated CGs. In the presence of excess factor protein templating and maintenance methylation restore the original chromatin structure **c**.

that, in active chromatin, histones H2A and H2B are less firmly bound to the nucleosome, and that it is the nucleosome itself which indicates which is active and which is inactive chromatin.

4.2 DNA methylation

Most minor bases are found in DNA in symmetrical sequences (e.g. G^{m}ATC or C^{m}CGG or, in vertebrates, simply ^{m}CG). Certainly in vertebrates, not all such sequences are methylated, and so DNA contains some methylated and some unmethylated CGs. Methylation of DNA is a post-replicative event in which a methyl group is transferred from S-adenosyl methionine to the acceptor base in DNA. When all the acceptor sequences are fully methylated (as in prokaryotes) the methyltransferase need show no preference, but, in eukaryotes, where the pattern of methylation is transferred from one generation to the next, some means is required to identify which CGs are to be methylated after replication. This is achieved by means of maintenance methylation (20).

A maintenance methylase recognizes the hemimethylated structure arising at replication of a pair of ^{m}CGs and acts to transfer a methyl group to the unmethylated cytosine (*Figure 6.3*). Part of the recognition signal is the presence of a methylcytosine on the parental strand. In the absence of such a methylated parental strand, no methylation of the daughter strand occurs.

If methylated bases are signals that affect the binding of chromatin proteins, then much of the need for protein templating described in the previous section becomes unnecessary.

5. Conclusions

This chapter has been concerned, largely, with the problems caused in eukaryotes by the association of the DNA with a large amount of protein. Although these proteins are important for packaging the DNA and for controlling its expression they also need to be synthesized, and yet at the same time it is not immediately clear what happens to the proteins whilst the DNA is replicating.

Heterochromatin is late replicating and this may reflect the time required to unravel such a structure to make it available to the appropriate enzymes. As adjacent replicons replicate together (Chapter 4, Section 1) it is clear that extensive regions of chromatin initiate replication together. As the replication fork moves along a section of chromatin, histone H1 is displaced and this may also be true for most transcription factors. The nucleosomes may transiently dissociate from the DNA and quickly reassociate with the leading strand. It is probable that the fork moves in a series of nucleosome-sized jerks along the DNA and this might account for the size of Okazaki pieces in eukaryotes (Chapter 1, Section 4).

As the replication fork passes, there is competition between various DNA binding proteins. If transcription factors are present they will bind in preference to histones but, in highly methylated, non-transcribed regions, histone H1 will bind and the region will rapidly become nuclease resistant and heterochromatic.

Much of this conclusion is speculative, but it provides a framework which is consistent with observations and on which further experiments can be based.

6. Further reading

Adams,R.L.P. and Burdon,R.H. (1986) *The Molecular Biology of DNA Methylation*. Springer Verlag.

Marzluff,W.F. and Pandey,N.B. (1988) Multiple Regulatory Steps Control Histone mRNA Concentration. *Trends in Biochem. Sci.*, **13**, 49.

Richmond,T.J., Finch,J.T., Rushton,B., Rhodes,D. and Klug,A. (1984) Structure of the Nucleosome Core Particle at 7Å Resolution. *Nature (London)*, **311**, 532.

Von Holt,C. (1985) Histones in Perspective. *BioEssays*, **3**, 120.

Weintraub,H. (1985) Assembly and Propagation of Repressed and Derepressed Chromosomal States. *Cell*, **42**, 705.

Wu,R.S., Panusz,H.T., Hatch,C.L. and Bonner,W.M. (1986) Histones and their Modifications. *CRC Crit. Rev. Biochem.*, **20**, 201.

7. References

1. Stein,G.S., Plumb,M.A., Stein,J.L., Marashi,F.F., Sierra,L.F. and Baumbach,L.L. (1984) *Recombinant DNA and Cell Proliferation*, Chapter 5, 107. Academic Press.
2. La Bella,R., Gallinari,P., McKinney,J. and Heintz,N. (1990) *Genes and Develop.*, **3**, 1982.
3. Russev and Hancock, (1982) *Proc. Natl. Acad. Sci. USA*, **79**, 3143.
4. Leffak,I.M. (1984) *Nature (London)*, **307**, 82.

5. Louters,L. and Chalkley,R. (1985) *Biochem.*, **24**, 3080.
6. Worcel,A., Han,S. and Wong,H.L. (1978) *Cell*, **15**, 969.
7. Jackson,V. and Chalkley,R. (1981) *Cell*, **23**, 121.
8. Dilworth,S.M., Black,S.J. and Laskey, R.A. (1987) *Cell*, **51**, 1009.
9. Seale,R.L. (1978) *Proc. Natl. Acad. Sci. USA*, **75**, 2717.
10. Weintraub,H. (1976) *Cell*, **9**, 419.
11. Sogo,J.M., Stahl,H., Koller,Th. and Knippers,R. (1986) *J. Mol. Biol.*, **189**, 189.
12. Bonne-Andrea,C., Wong,M.L. and Alberts,B.M. (1990) *Nature (London)*, **342**, 719.
13. Riley,D and Weintraub,H. (1979) *Proc. Natl. Acad. Sci. USA*, **76**, 328.
14. Seidman, M.M., Levine,A.J. and Weintraub,H. (1979) *Cell*, **18**, 439.
15. Workman,J.L., Abmayr,S.M., Cromlish,W.A. and Roeder,R.G. (1988) *Cell*, **55**, 211.
16. Solomon,M.J. and Varshavsky,A. (1987) *Mol. Cell. Biol.*, **7**, 3822.
17. Brown,D.D. (1984) *Cell*, **37**, 359.
18. Weintraub,H. (1979) *Nucl. Acids Res.*, **7**, 781.
19. Kumar,S. and Leffak,M. (1986) *Biochem.*, **25**, 2055.
20. Adams,R.L.P. (1990) *Biochem. J.*, **265**, 309.

Glossary

Amplification: an increase in the copy number of a gene or series of adjacent genes.

Antisense RNA: RNA with a sequence complementary to the normal transcript, e.g. mRNA.

Autonomous replication: replication not under the control of a host chromosomal origin, i.e. as a plasmid.

Catenated: joined together, either end to end or as the links in a chain.

D-loop structure: triplex DNA structure associated with some origins of replication where one of the parent DNA strands is displaced by a newly synthesized third strand that is hydrogen bonded to the other parental strand.

Distributive: dissociation of the enzyme (polymerase) from the substrate (primer) after each addition step.

***dna* mutants:** mutants (usually temperature sensitive) which are defective in some stage of DNA replication.

Epigenetic information: information passed from generation to generation, but not encoded in the sequence of bases in the DNA.

Fibre autoradiography: autoradiography of tritium-labelled fibres of DNA stretched out on a slide.

Gap: a region of duplex DNA in which several nucleotides are missing in one strand.

Hairpin structure: a palindromic sequence (of RNA or DNA) that can fold back to hydridize with itself—usually with a short region of non-palindromic nucleic acid separating the two halves of the palindrome.

Histone octamer: the eight histone molecules (two each of H2A, H2B, H3, and H4) that make up the nucleosome core.

Incompatibility: inability of two phages or plasmids to coexist in the same host cell.

***in vitro* complementation assay:** an *in vitro* assay in which the deficiency of a component can be complemented by its supply from another source. This enables the amount of the component to be quantitated.

Leading strand: The daughter strand which is made first (and continuously) at the replication fork.

Lagging strand: the daughter strand which is made later (and discontinuously) at the replication fork.

Linking number: specifies the number of times that the strands of DNA are wound around each other. For a circular molecule the linking number (α) is given by: $\alpha = \beta + \tau$; where β is the number of Watson–Crick turns and τ is number and sense of the superhelical turns.

Minichromosome: the chromosome formed of plasmid or viral DNA and nucleosomes. Usually refers to the SV40 minichromosome.

Nick: a break (usually with a 3′ OH) in the phosphodiester backbone of DNA. No phosphate or deoxyribose is missing.

Nuclear matrix: the nucleoprotein complex that remains when dehistonized chromatin is treated with nucleases. It is believed to form the structural backbone to the chromosomes.

Polymerase chain reaction (PCR): a reaction in which primers are annealed to DNA and extended by the *Taq* polymerase (from *Thermus aquaticus*). Multiple rounds of annealing and extension are possible without further addition of enzyme as as result of its thermal stability. The use of two contrary facing primers allows the amplification of the intervening region.

Preprimosome: A complex of proteins associated with, and translocating along, a prokaryotic chromosome. It may play a part in unwinding the DNA at the replication fork, and is important in primosome formation.

Primer: an oligoribo- or deoxyribonucleotide with a 3′ OH end upon which a nucleic acid chain may grow.

Primosome: a complex of the preprimosome and primase, important in primer formation.

Processive: addition of multiple monomer units by an enzyme that does not involve dissociation of the enzyme:substrate complex at each stage.

Proofreading: checking each nucleotide addition for complementarity to the template before proceeding to the next addition.

Redundant sequence: a sequence of DNA that is present in more than one copy, albeit not necessarily in the same orientation.

Replicative form (RF): the form in which viral DNA replicates in the cell i.e. a double-stranded form.

Replicon: that DNA replicated from a single origin.

Retrograde synthesis: synthesis (of DNA) in a direction opposite to that in which the replication fork is moving.

Template: the strand (in this case, of DNA) that dictates the order of addition of the incoming nucleotides during replication.

Transcription factor: a protein that (usually) binds to the DNA to facilitate gene transcription.

Index